The Social Causes of Environmental Destruction in Latin America

Linking Levels of Analysis
Emilio F. Moran, Series Editor

Covering Ground: Common Water Management and the State in the Peruvian Highlands
David W. Guillet

The Coca Boom and Rural Social Change in Bolivia
Harry Sanabria

Diagnosing America: Anthropology and Public Engagement
Shepard Forman, editor

The Social Causes of Environmental Destruction in Latin America
Michael Painter and William H. Durham, editors

Culture and Global Change: Social Perceptions of Deforestation in the Lacandona Rain Forest in Mexico
Lourdes Arizpe, Fernanda Paz, and Margarita Velázquez

The Social Causes of Environmental Destruction in Latin America

edited by
Michael Painter and
William H. Durham

THE UNIVERSITY OF MICHIGAN PRESS

Ann Arbor

Published in the United States of America by
The University of Michigan Press
Manufactured in the United States of America
♾Printed on acid-free paper

1998 1997 1996 4 3 2

A CIP catalogue record for this book is available from the British Library.

Library of Congress Cataloging-in-Publication Data

The social causes of environmental destruction in Latin America /
edited by Michael Painter and William H. Durham.
p. cm.—(Linking levels of analysis)
Includes bibliographical references and index.
ISBN 0-472-09560-9 (alk. paper). — ISBN 0-472-06560-2 (pbk. : alk. paper)
1. Environmental degradation—Latin America. 2. Deforestation—Environmental aspects—Latin America. I. Painter, Michael. II. Durham, William H. III. Series.
GE160.L29S63 1995
363.7'01'098—dc20 94-34146
CIP

Acknowledgments

This volume began as a panel at the Annual Meetings of the American Anthropological Association in Washington, D.C., in 1989. In the intervening years, the chapters have been revised, expanded, and sharpened as the manuscript worked its way through several revisions, all under the caring, prodding supervision of Emilio Moran, editor of the series on Linking Levels of Analysis. We would like to express our appreciation to Emilio, to the editorial staff at the University of Michigan Press, and most of all, to the authors of the chapters, who met every challenge with both professionalism and good will. The final product is a convincing testimonial of their dedication.

Preparation of the manuscript was supported in part by the SARSA Cooperative Agreement between the Institute for Development Anthropology, Clark University, Virginia Polytechnic Institute and State University, and the U.S. Agency for International Development. The lion's share of the work in this process was undertaken by the editorial staff of the Institute for Development Anthropology and, in particular, by Sylvia Horowitz, whose careful editing and good humor were appreciated by all. Last but not least, the editors express their gratitude for this opportunity to work together across many miles.

Series Introduction

The series Linking Levels of Analysis focuses on studies that deal with the relationships between local-level systems and larger, more inclusive systems. While we know a great deal about how local and larger systems operate, we know a great deal less about how these levels articulate with each other. It is this kind of research, in all its variety, that Linking Levels of Analysis is designed to publish. Works should contribute to the theoretical understanding of such articulations, create or refine methods appropriate to interlevel analysis, and represent substantive contributions to the social sciences.

The volume before you, *The Social Causes of Environmental Destruction in Latin America,* is the first edited volume in the series. It is no mean task to accomplish the kind of interlevel analysis we seek to make available in a single-author work. To bring together yet a number of authors and provide a coherent and consistent interlevel analysis is indeed a major feat. The editors, Michael Painter and William H. Durham, have succeeded where many would have failed. They organized one of the most exciting symposia in years at the annual meeting of the American Anthropological Association and worked with the best of those papers to produce a superb social analysis of the processes behind environmental destruction in Latin America.

The authors show the linkage between the processes of deforestation in Central and South America, and they account for the international pressures experienced by particular nations. However, the authors also show that the internal processes derived from class oppression, urban versus rural preferences, and land tenure and tax law choices are no less significant. Most readers will come to this book with the view that the fate of these nations is largely a product of their dependency on export production and their "periphery" status in a world system. The authors of these chapters will convince readers that the processes are far more complex, involving appropriation of land and other natural resources associated with particular patterns of social class formation. The chapters also show that solutions to environmental degradation may need to target areas other than where the environmental degradation is taking place. Unequal exchange relations between different classes lead to overuse of resources elsewhere, as people relocate to escape inequities in areas of origin. Most colonization projects in the Amazon are a product not of overpopulation or competition for the resources of the forest but result from the political unwillingness of landed elites to redistribute land in areas of latifundia. In turn, they export the same inequities to the frontier, where such differentials have been far less. In short, this volume shows how conflicting and unequal claims to land between linked political and economic actors are largely responsible for environmental destruction in Latin America. It is the kinds of linkages and relationships people have with each other that lead to environmentally sound or destructive behavior. Thus, the search for environmentally appropriate behavior must address the oppressive inequities found throughout the region in access to resources.

It is my hope that this volume will not only be a major contribution to our understanding of these processes but that it will inspire future contributions to the series. Please contact the series editor or other members of the editorial board to discuss your proposed work and our possible interest in publishing it.

Contents

Conclusion

Introduction: Anthropological Perspectives on Environmental Destruction

Michael Painter

Anthropological Studies of Production and Environment

Anthropology has offered numerous insights into how people perceive and use the environment, and anthropologists have substantially augmented public awareness of the environmental problems that face contemporary societies. Yet, despite a tradition of studying the relationship between the social lives of human populations and their physical surroundings, environmental destruction and what to do about it have not been defined as anthropological research issues by the discipline. As a result, discussions about how humans use land and other natural resources are replete with references to culture, indigenous knowledge systems, institutions that define and regulate resource access, and other concepts substantially or totally drawn from anthropological research. Interpreting the significance of these concepts for environmental policy and political action, however, has been largely left to nonanthropologists. For example, the *Anthropology Newsletter*, the bulletin of the American Anthropological Association, complains:

> Two cover stories on the environment (*Time Magazine*, "Torching the Amazon: Can the Rainforest be Saved?" September 18, 1989; and *Scientific American*, Special Issue: "Managing Planet Earth," September 1989) are peppered with anthropology-relevant words and phrases: "culture shift," "traditional culture," "pronatalist culture," "peasants," "Neolithic," "humankind," "desertification," and so on. Yet **anthropologists** are neither quoted or cited

anywhere in the magazines [emphasis retained from the original text] (*AN* 30 [8]:2 [Nov. 1989]).

One reason that anthropologists do not receive the attention we think we deserve is that discussions of anthropological theory have often denied or overlooked the relationship between political interests and scholarly theory (e.g., Ehlers 1990), despite the strong relationship that has existed between the two in important areas of anthropological research. Speck (1915), for example, investigated the relationship between property rights and natural resource husbandry among Native American populations of North America around the turn of the century. His discussions of how these populations defined hunting territories over which kin groups exercised exclusive rights, and pursued resource management strategies intended to prevent the depletion of fish and game, are remembered for their importance in refuting the evolutionary model of social organization proposed by Lewis Henry Morgan. That his data were gathered as part of an effort to resist the Canadian government—which sought to support private groups hoping to take over Native American lands—is often forgotten (Feit 1986).[1]

Since the time of Speck, the relationship between people and the physical environment has been an important issue in anthropological research. Kroeber (1939) and Wissler (1938) examined the geographic distribution of cultural traits and developed the concept of culture areas. However, anthropologists were concerned that this not be linked to environmental determinist arguments that sought to explain and justify the domination of much of the world by people from the temperate latitudes of the northern hemisphere through references to climatic factors that supposedly favored a particularly vigorous and industrious population. This discouraged an examination of humans' relationship to the environment as an interactive process and emphasized humans' transformation and appropriation of their surroundings, with the environment accorded only a passive "possibilist" role (Herskovits 1951).

The passivity accorded the environmental side of the human-environmental relationship reached its extreme in the unilineal evolutionism of White (1949), which defined cultural evolution in terms of human capacity to capture energy from the environment. White argued that "environmental factors may . . . legitimately be considered a constant and as such omitted from our consideration" (1949:199). Conceived in an era when human capacity to use technology to capture energy from new sources appeared unlimited, White's model considered only the total amount of energy captured. White did not consider the efficiency with which energy was used, nor how populations might respond to resource scarcity. It is telling that the major critique of unilinear evolution (Steward 1955) focused on demonstrating that human responses to a given set of conditions are more variable and multifaceted than allowed for by White, while the ideas of the passivity of the environment and the capacity of humans to extract ever-increasing quantities of energy from it were not critically examined.

Steward's approach turned on the ways in which people use technology to exploit or transform the environment through production. Production was a technical activity mediating adaptation to the environment, and the way people used natural resources to produce was a function of technological advancement. Social organization could change according to the need to manage a particular productive activity at a given level of technology, and within the passive constraints imposed by the environment. Reacting to anthropology's tendency to link history to culture history, which was rooted in the idea that culture traits diffused from one population to another and embodied a rejection of evolutionary models of social change, Steward largely rejected a historical perspective. The anthropological view of history, he felt, trapped one in particularistic explanations of change that precluded linking specific cases to general processes (1955:78–97, 208–209).

Thus, in *The People of Puerto Rico* (1956), which was the field demonstration of research methods behind multilinear evolution, Steward argued that changes in production practices and social organization should be explained in terms of environment and technology:

> The method of cultural ecology requires first an examination of the relationship of technology, or productive processes, to the environment. . . . Next it is necessary to ascertain the cultural and environmental factors which cause certain of these to be selected by the new area. Finally, the modifications in the productive processes in a new environment must be analyzed and their effects upon other aspects of culture determined. When a particular exploitative technology such as farming is introduced to a new region, the local soils, topography, rainfall, climate and other environmental factors will usually

> require modifications in methods of production and in the related patterns of marketing, land tenure, co-operation, and settlement patterns. (Steward 1956:15)

Steward's cultural ecology method crystalized problems that have continued to influence anthropological approaches to how humans appropriate natural resources. First, because his notion of production focused on the technical relationship between the immediate users of a technology and the environment, contextualizing local situations with respect to regional, national, and international processes presented a continuing difficulty. Second, Steward's rejection of the idiosyncratic notion of history that had developed in anthropology extended to a rejection of history generally. This rendered cultural ecology essentially synchronic in its perspective, and unable to explain very well how a local situation had come to be.

Many anthropologists continued to treat production as a technical process and to pay little attention to the historical context in which it occurred. Successive incremental refinements in techniques for measuring aspects of production permitted analyses to be increasingly fine-tuned. The resulting models were descriptive rather than explanatory, however. Moreover, the descriptions were based on normative statements about production and the environment. Falsely emphasizing the harmony between natural resources and human objectives, they portrayed production systems in terms of a stability that often bore little resemblance to actual conditions.

Anthropologists also persisted in attempting to divorce scientific endeavor from the political context in which it occurred, but their descriptive models lent themselves to politically charged decisions about land use policy and practice. For example, Hjort (1982), criticizes ecological models of pastoral production systems in Africa for their normative descriptive quality. He points out that the "normal" condition is one of extreme seasonal and annual variation in rainfall and vegetation, and constant alternation between periods of scarcity and abundance (1982:15). In addition, ecological models fail to consider the chronic land grabbing to which pastoral populations have been subjected. This "normal" condition for contemporary pastoral peoples has resulted in losses of their land to government officials and other largely urban-based elites, reducing the geographic and biological diversity to which they have access (Hjort 1982:20; Little 1985).

Hjort continues that treating production as if it were part of a self-regulating system that sometimes gets out of balance for technical reasons may identify the symptoms of environmental degradation, but avoids addressing the real causes, such as inequities in access to critical resources. The implications of such an approach may be disastrous for the pastoral population itself. Assuming that a system tends to be balanced and self-regulating makes it easier to demonstrate the rationality of production. But, since most modern systems are not isolated, as would be necessary for the posited self-regulating mechanisms to function, waiting for them to balance themselves without addressing such issues as external competition for land and other unplanned changes may be as fatal for pastoral people as a direct assault on their way of life (Hjort 1982:22–23). Treating technical aspects of production, without considering the historical issue of how a production system came to be or without contextualizing it in terms of competing interests in the same resources, provides information that may be manipulated in various ways to the detriment of the politically weakest people with an interest in an area. For example, Hjort sees models that treat systems as self-regulating as promoting passivity in the face of crises in pastoral production. At the same time, as Horowitz (1986) and Sandford (1983) point out, focusing on the technical aspects of how a particular herd management practice degrades the environment, without placing the practice in a broader context, invites the interpretation that pastoralists themselves are destructive. Such arguments have frequently been used to justify forced sedentarization and various schemes to separate herders from their land.[2]

Despite the difficulties of cultural ecology as conceived by Steward and his descendants in anthropology, *The People of Puerto Rico* also provided the forum for a different approach to the study of human production that has come to offer more fruitful insights into how and why environmental degradation occurs.[3] While Steward accorded history little importance in constructing explanations of production and environmental change, the volume contains an extended historical discussion (Staff 1956) that is remarkable on two counts. The fact that it is contained in a volume whose introduction, describing the method and theory on which the study was based, accords history no importance is suggestive of the limitations that the researchers found in cultural ecology's utility for interpreting the significance of the rich body of data collected by the study. Roseberry (1978) observes:

> While the investigators mention the importance of "the environment" and Steward mentions the importance of "history," the two statements of method are incompatible. The disagreements are stated in polite academic language. But any social scientist who has disagreed with a teacher or a colleague in the course of a seminar or a joint project should recognize what is really being said. (1978:31)

At the time Steward wrote, the most common use of historical materials by anthropologists was to justify the reification of cultural particularities. Steward's rejection of the importance of history was a reaction to this tendency. The remarkably extensive historical discussion in *The People of Puerto Rico* departs from conventional discussions of history by anthropologists. It explains the variation in Puerto Rican subcultures in terms of local responses to the demands exerted by international commodity markets and the resulting global reorganization of production relations (Staff 1956:32).

Despite the significance of this integrative focus, it was not without its own shortcomings. For example, the study focused on agricultural production defined in a narrow, empirical sense, and while it makes many references to external forces, capitalism is not explicitly discussed (Roseberry 1978:33). As a result, phenomena such as migration to the U.S. mainland, which were locally important but owed their origin and perpetuation to forces that were at work internationally, were not considered:

> The Puerto Rico project, in its concentration on agriculture, failed to come to grips with the political and economic forces that established that agriculture in the first place, and that were already at work in "Operation Bootstrap" to transform the agricultural island into an industrial service station. We did not understand the ways in which island institutions, supposedly "national" but actually interlocked with mainland economies and politics, were battlegrounds for diverse contending interests. (Wolf 1990:589)

Nonetheless, through its carefully crafted case studies of different agricultural production regimes and the historical discussion that joined them, *The People of Puerto Rico* demonstrated that patterns of production and resource use are more than the outcome of the application of a particular level of technology to a set of environmental constraints.

Rather, the definition of what natural resources are important in a particular time and place, how and to what end they are exploited, and the overall relationship of the population to the physical environment flow from the social relations that define and regulate differential access to those natural resources.

The resolution of subsequent lacunae affecting parts of sociocultural anthropology was eased by the theoretical and methodological ground broken by the Puerto Rican research. For example, in the 1960s economic anthropology became mired in circular discussions about the appropriateness of applying formal economic analysis to non-Western societies to such an extent that some scholars thought it was reaching the limits of its utility as an area of inquiry. The solution offered was that economic anthropologists should use ecological models as a means of understanding non-Western production systems (e.g., Vayda 1967, 1969). Through its use of historical analysis to show how the diverse situations of sugar plantation workers, independent coffee growers, and diversified family farmers are linked through the reorganization and subordination of production arrangements to supply capitalist markets, *The People of Puerto Rico* showed how the environment, as it is commonly perceived, and technology are socially constructed outcomes of production relations between people, rather than empirical givens on which to base social analysis. Subsequent anthropologists (e.g., Cook 1973, 1974) treated production as a social process occurring between humans, for the satisfaction of sometimes conflicting needs, rather than as a technical response by an undifferentiated population to conditions in the physical environment.

The Social Causes of Environmental Degradation

By examining resource use in terms of social purpose, anthropological research has shown that the people who control access to natural resources and the institutional arrangements through which that control is mediated shape the sorts of resource management practices that will be followed within the constraints of the physical environment. Rather than treating the environment as a passive entity that imposes broad limits on human activity, such an approach focuses on the dynamic relationship between human productive activity and the physical resource base. The nature of this activity is shaped by the social determination of what constitute critical natural resources at a particular time and place, the distribution of access to those resources, and the nature of the institutional arrangements mediating that access.

Wolf (1972) has described such an approach to the study of human production as "political ecology," and a number of anthropologists have applied the notion productively to the study of environmental degradation. Hjort (1982) made explicit use of the term in his critique of models of pastoral land use, cited above, as did Schmink and Wood (1987) in their suggestion of six elements critical to the understanding of settlement and environmental destruction affecting the Amazonian lowlands. These elements include: (1) the nature of production in a region, and, in particular, the degree to which it is oriented toward simple reproduction or capital accumulation; (2) the class structure of the society to which the region in question belongs and the lines of conflict over access to productive resources; (3) the extent and kinds of market relations in which producers are involved, and the mechanisms whereby production beyond that needed to satisfy consumption requirements is extracted as surplus; (4) the role of the state in defining and executing policies that favor the interests of certain classes of resource users over others; (5) international interests, such as donor agencies or private investors, that may support particular patterns of resource use; and (6) the ideology that orients resource use—for example, the position that rapid economic growth is the best way to address social and environmental problems—and what groups benefit from that ideology.

Research based on these points that has examined environmental destruction in Latin America has clarified several issues that are central to any plan of action seeking to reverse or ameliorate the problem. First, environmental destruction associated with the production systems of smallholding farmers is a consequence of their impoverishment, either absolute or relative to other social classes. This impoverishment often has occurred together with loss of land and subjection to violence at the hands of wealthy individual and corporate interests engaged in land speculation and of state authorities (e.g., Collins 1986). Second, while easier access to smallholders has meant that the environmental destruction associated with their production systems has received the most attention, much more land has been degraded by the activities of wealthy individual and corporate interests. Generally, large-scale enterprises that have acted destructively have been granted land on concessionary terms by the state exercising sovereignty over the area in which they operate. This allows them to treat land as a low-cost input, and makes it more economical to move elsewhere after the environment is degraded than it would be to try to conserve (e.g., Bakx 1987; Binswanger 1991). Third, the same policies and practices that result in wealthy interests receiving land on concessionary terms are responsible for the impoverishment of smallholders, because they institutionalize and exacerbate unequal access to resources. Thus, the crucial issue underlying environmental destruction in Latin America is gross inequity in access to resources (e.g., Painter 1990).

The chapters included in this volume draw on data gathered in Central and South America to illustrate the social and economic processes that underlie environmental degradation and underdevelopment. Individually and collectively, the chapters argue that addressing these issues means dealing explicitly with the fact that both processes occur in complex societies joined to one another by a global economy and ecology, and characterized by social classes with fundamentally divergent interests. Which members of a population have their access to productive resources enhanced, and which have it restricted are of profound importance in shaping patterns of economic growth and natural resource use.

The chapters in this volume are divided into South American and Central American cases. Although this arrangement is, in many regards, arbitrary—issues such as deforestation clearly cut across the two regions—it is intended to highlight broad regional differences in the social context of environmental degradation. In South America, the competition for resources among social classes is normally conceived by foreign and national observers alike as having both national and international dimensions, while in Central America, discussion centers much more around international, particularly U.S., interests. In South America, environmental degradation is much more commonly associated with specific policies and programs—the implementation of the agrarian reform and the expansion of lowland commercial agriculture in Bolivia, or the construction of the Carretera Marginal de la Selva and the implementation of "special projects" for tropical forest development in Peru, for example. Clearly, important international interests were at work—U.S. bilateral assistance financed the initial expansion of commercial agriculture in eastern Bolivia. However, these were consistently treated in tandem with national class interests, like the assertion of control by a national bourgeoisie over the country's mining industry and its capture of the state as a source of employment. In Central America, environmental degradation by and large has been treated as a by-product of U.S. geopolitical machinations and dependence on international markets.

The contrast in how environmental issues are framed has its roots in historical differences between the two regions. Central American states fought no wars with Spain, and independence did not carry the trappings of political struggle usually associated with it. The major conflict enveloping the region at the time was resistance to incorporation into the Mexican empire that Iturbide sought to establish. The formation of the Central American Federal Republic, a liberal initiative based on secularism and free trade, proved inadequate to consolidate liberal political power in the face of opposition from the peasantry, the Church, and other landed interests. During the two decades following political independence, the status of what are today's Central American nations changed from states to provinces and, finally, in 1838, to republics. They were, in Dunkerley's words, "parts that could not make a whole" (1987:4).

The region lacked the kinds of resources and industries, like guano, minerals, and sugarcane, that drove the internal processes of class formation and conflict, and the diverse relations that existed between national entrepreneurial classes and international markets in South America. Mineral resources were few, and the major agricultural export at independence, cochineal, was in decline. The coffee and banana production regimens that we currently associate with Central America did not emerge and begin to consolidate themselves until around the turn of the century—later in some areas—and cattle ranching to supply export and domestic urban beef markets is a decidedly post–World War II phenomenon.

In such a context, internal processes of political struggle and class formation were often submerged in and subordinated to the geopolitical interests of external powers seeking to exert control over the region. These revolved around Central America's strategic importance as a passage between the Atlantic and Pacific oceans, and the influence that a power exercising hegemony over the Caribbean basin could exert over U.S. commercial expansion through the Mississippi River system. During most of the nineteenth century, Great Britain was the dominant foreign power. U.S. influence was expanding, however, and by the twentieth century was clearly manifest in a growing political and military intervention and commercial domination that ultimately vitiated the national and regional political sovereignty of the Central American nations.

Given this background, it is not difficult to see how explanations of social and economic change or environmental degradation in Central America often begin and end in the United States. Unfortunately, while such accounts are useful for emphasizing the interconnectedness of socioeconomic processes that we observe in different parts of the world, they easily degenerate into functionalist circularity whereby everything is explained in terms of how it serves the interests of a fetishized and monolithic construction of the capitalist world economy with vague reference to well-known (and less well-read) dependency theorists and their intellectual descendants (see Brenner 1977; Mintz 1977). At an empirical level, such a line of argument is unappealing because it ignores the tremendous diversity of local initiatives and responses to external pressures, and because it falsely implies that those external pressures are exerted in a coherent fashion from an internally homogeneous source.

A recent and particularly well-known example of this line of argument is the "hamburger thesis," which seeks to explain deforestation in Central America in terms of the expansion of the region's export beef industry in response to the growing U.S. appetite for fast-food hamburgers. Each of the Central American articles provides a different and powerful refutation of this kind of reasoning. The most direct is the contribution by Marc Edelman, which draws on Costa Rican data to question the efficacy of the hamburger argument. Edelman first challenges it on empirical grounds, pointing out that the coincidence between the expansion of beef exports to the United States and deforestation was a conjunctural event, and that, while stagnation of the export beef industry has coincided with a reduced rate of conversion of forest to pasture, deforestation has continued. A number of factors internal to Costa Rica have been instrumental in shaping the behavior of cattle ranchers, including a decline in per capita beef consumption while aggregate domestic demand for beef has increased, and land tenure and tax laws that encourage extensive land use for ranching. Edelman points out that the purpose of converting forest to pasture is not to increase the amount of grassland, but to make a profit, and that several institutional arrangements in Costa Rica favor the making of profit through the appropriation of natural resources or natural processes so that these do not appear as costs in the production process (the so-called subsidy from nature). In this light, we see that the pressures on Costa Rica's tropical forest resources are unlikely to diminish as a result of declines in the U.S. demand for beef.

While not addressing the hamburger thesis directly, Susan Stonich echoes many of the points made by Edelman in her discussion of southern Honduras. She places the expansion of export beef production in the context of the histories of other export industries, including cotton, shrimp, and melons. She demonstrates that it is not necessarily export production itself that underlies environmental degradation, but the superimposition of production for export markets onto internal processes involving the appropriation of land and other natural resources associated with particular patterns of social class formation. For example, while environmental impacts are clearly associated with the establishment of commercial shrimp ponds in mangrove swamps, the long-term environmental threat may come from the large numbers of rural people who have been displaced because their land has been appropriated and their artisanal fishing and gathering activities disrupted. The shrimp industry provides few employment opportunities to absorb this displaced population, so that most are left to make a living as best they can on the declining area of unappropriated land.

Thus, export production contributes to environmental degradation by aiding and abetting internal processes of capital accumulation whereby the wealthy appropriate natural resources without being accountable for the social or environmental consequences of their actions. Such actions displace large numbers of poor people who, denied access to the means of earning a living within those areas of the economy where accumulation is occurring, have little recourse but to behave in a predatory fashion toward an increasingly impoverished, unappropriated resource base. The point has methodological and political implications for those concerned about environmental degradation as a global process. Regarding the former, Stonich's article is a strong warning about the kinds of critical issues that those relying on vulgar versions of dependency or world systems theory are likely to overlook. Politically, Stonich's contribution highlights the problem with the "small-is-beautiful" version of agrarian populism to which many who are concerned about the environment and/or the plight of the rural poor have been susceptible.[4] For many countries characterized by some combination of natural resource constraints and a small population, export production may be the only choice for generating rural income. In such contexts, the appropriate focus for political action may be the social arrangements under which that production takes place, rather than the fact of export production itself.

In his discussion of settlement and deforestation in the Petén of

Guatemala, Norman Schwartz seeks "to understand how what no one wanted . . . but no one halted came about." According to his account, all sectors of Guatemalan society with an interest in the Petén agreed on the importance of avoiding uncontrolled spontaneous settlement and deforestation. However, the social context did not enable people to forgo short-term economic interests in pursuit of medium- and long-term advantages that would also benefit them. Schwartz's chapter emphasizes how class relations within Guatemala drive the settlement and deforestation of the Petén. Export markets for products such as mahogany accelerate the process, but internal class relations are the driving force. Schwartz also makes an important point in observing that, while the environmental problem is physically located in the Petén, the social relations defining access to land and patterns of capital accumulation are national in scope. Thus, efforts to halt or slow settlement and deforestation are unlikely to be successful if they are confined to the Petén.

The South American cases in the volume contain themes similar to those developed in the discussions of Central America. One theme continued into the second part of the book is that the solutions to environmental degradation do not necessarily lie in the area where the degradation is occurring. My chapter on central Bolivia, for example, has two clear cases of this. First, the environmental degradation afflicting the semiarid upland valleys is a product of unequal exchange relations between agriculture there and Bolivia's mining and industrial agriculture sectors. The resulting decline in rural incomes has caused rural families to be heavily dependent on off-farm work, which results in chronic labor scarcity and the progressive abandonment of medium- and long-term land management practices. Second, the impoverishment of the area has led to long-term and permanent out-migration. The largest portion of these migrants has gone to Bolivia's cities, particularly Cochabamba, La Paz, and Santa Cruz. A significant number of people have also migrated to lowland colonization areas, including northern Santa Cruz Department and the Chapare region of eastern Cochabamba Department. Because migration into the Chapare was driven by land scarcity in upland areas, rather than any particular opportunity to earn a stable living through farming in the tropical lowlands, large areas were cleared for agriculture that were unsuited for that purpose. The unfortunate boom in the international demand for cocaine during the late 1970s and the 1980s accelerated migration to the Chapare. Current attempts to ameliorate the environmental degradation associated with settlement of the Chapare and to provide farmers with alternatives to producing coca leaf for manufacturing cocaine depend on resolving the land crisis in population-expelling areas, and on establishing favorable market relations that allow Chapare farmers to sell their produce profitably.

James Jones echoes important elements of this theme from a different perspective. He relates environmental problems in Bolivia's Beni region to ethnic relations between native people there and Bolivians involved in different periods in rubber extraction, the peltry trade, ranching, and lumbering. Frequently, the loci of political and economic interests that have shaped interethnic relations have been located outside of the Beni, or, in the case of rubber, outside of Bolivia altogether. The monopolization of power to decide how land will be used by people who do not live there and who do not depend on the use values of the products of the land for their livelihood is a key cause of environmental degradation in the area. It is also the issue that has provided the major stimulus for the political mobilization of native people in the Beni to assert their rights over the land. Jones also notes that the continuing northward expansion of settlers from the Chapare into the Isiboro-Sécure area is creating an explosive situation between two populations of poor people. This is a particularly unfortunate example of what happens when the physical location of land differs from the social location of the struggle over how that land is to be used.

Eduardo Bedoya's contribution shifts our attention from Bolivia to Peru, where he looks at how broad social struggles create family-level pressures to use land in particular ways. Bedoya compares the rates of deforestation associated with agriculture in different contexts: settlers in Peru's Upper Huallaga valley, and three native populations with different relations to settlers and to commodity markets. As one would expect, settlers in the Upper Huallaga are the most destructive of tropical forest. This is related to the hostile economic climate for smallholders in Peru generally, and to how upland social and economic problems drove settlement in the Upper Huallaga, in ways similar to the Chapare, so that considerable farming takes place on land that is poorly suited for agriculture. At the same time, there is considerable differentiation within the Upper Huallaga settler population with respect to the rate of deforestation associated with their agricultural practices according to such factors as the tenure arrangements through which they hold land. Bedoya's comparison of the production systems of three native populations—the Amarakaeri, the Machiguenga, and the Ashaninka—is also revealing.

While all three production systems are less destructive of tropical forest than those employed by Upper Huallaga settlers, the deforestation rate increases as the populations' land bases are depleted by incursions of settlers and the need to produce for commodity markets in order to satisfy the demand for cash.

Directions for the Future

Collectively, the contributors to this volume provide convincing documentation of the idea that relations between people with conflicting and unequal claims to land are primarily responsible for environmental destruction in Latin America. They also show the richness of anthropological tools and concepts, sometimes applied in unconventional ways, for illuminating the mechanisms whereby social relations drive environmental degradation. In addition, they point to important issues for future discussion and research on how we think about and how we attempt to address environmental degradation.

Each case demonstrates that how a particular population uses natural resources is as much an outcome of the kinds of relationships that people have with one another as it is a cause of environmental destruction. Thus, a search for forms of production that are more sustainable must begin by addressing the relationships that allow smallholders to lose access to land suited to farming, or that allow access to the privileged under conditions that offer no incentives to conserve. Seen in this light, advocacy of social equity and support for popular movements that seek to achieve equity is not demagoguery or an idealistic quest, but an ecological imperative.

Recognition of this point has important implications for environmental conservation. For example, the establishment of reserve areas has been a central part of the response to the massive environmental destruction occurring in South America. These reserves conform to several organizational models including parks, reservations for indigenous populations, controlled-use zones (e.g., extractive reserves), and biological protection areas. While some reserves promise a measure of environmental protection, they frequently are "paper parks," sites of conflict between native peoples and settlers, and sources of resentment because they offend concepts of national sovereignty. Projects to establish reserve areas often suffer from two fundamental problems: (1) they begin as if the planned reserves were isolated from the social and economic conditions that drive the environmental destruction motivating their establishment, and (2) they incorporate the creation and maintenance of such isolation as an explicit or implicit goal. To the extent that they grant preferred access to some (e.g., ecotourists) while restricting the access of people who depend on the land to survive (e.g., limiting productive activities to external notions of traditional hunting and gathering), reserve areas may become an additional constraint on access to land, exacerbating conflicts over the use of key resources, and accelerating the environmental degradation they seek to halt. It is worth remembering a point that Jones develops elsewhere (Jones 1991): one of the critical events contributing to the political mobilization that launched the March for Dignity—a march by native people of the eastern lowlands from the city of Trinidad to La Paz—was a "debt-for-nature swap" arranged among the Bolivian government, international banks, and Conservation International, which defined a conservation-oriented land use plan for the Chimán Forest with no participation from the native people living there.

A second applied task for anthropologists is the education of populations in the developed countries about their own relationship to the social and economic processes driving environmental degradation elsewhere. For example, Edelman notes that one problem of the "hamburger thesis" is that, by obscuring the mechanisms that induce deforestation in Costa Rica, it can lead to inappropriate political action (e.g., boycotting Burger King) by well-meaning people trying to contribute to a solution. A lot is contained in this observation that Edelman cannot develop in the context of his contribution to this book. One obvious issue is education—helping people to understand that, while we all are interconnected, the connections are not necessarily simple or direct, and helping them to learn how to discern what these connections are.

Such an educational effort opens up a broader questioning of our relations to one another; it should contribute to a perspective with more profound implications for changing behavior relating to our use of natural resources. If, for example, through a critical examination of the "hamburger thesis" we begin to appreciate the difference between consumer choice (e.g., trying to save the rainforest by eating chicken nuggets instead of hamburgers) and political action (e.g., promoting equitable access to resources for people whose lives depend on them), we may come much closer to living in harmony with other organisms on the planet.

Finally, the authors who have contributed to this volume have provided directions for thinking about how humans as a species relate to other species and to the physical resources that sustain all of us. The growing sense that our consumption requirements may exceed the availability of the resources needed to satisfy them, and that the burdens that this imposes are not likely to be shared any more equitably than the resources have been, has inspired a number of analyses of our notions of environment in anthropology (e.g., Vessuri 1986) and in other disciplines (e.g., Leff 1986). More broadly, many have begun to examine critically our concepts of political economy to see if, in discussing how to reorder the relations of production between people, we have placed too much emphasis on the transformative power of that production (e.g., Benton 1989; Grundmann 1991). Ultimately, how we answer may have much to do with the extent to which we see people, with our technologies, social classes, and specialized divisions of labor, as part of nature and a product of it, and to what extent we place ourselves outside of something that we objectify as nature (Schmidt 1971). Thus, the critical research task facing us is to define, with a rigor that we have not yet achieved, the relationships between what we commonly experience as the social and the physical domains of our lives. This is a task in which anthropologists have been engaged for some time, and to which we are well equipped to contribute.

NOTES

1. Foreshadowing arguments by Hardin (1968) and others, the Canadian government argued that Native Americans were profligate resource users because, in the absence of exclusively held (i.e., private) property, individuals were not accountable for how they used natural resources and had no incentive to behave in a conserving manner.

2. Natural scientists may be less reticent about discussing the limitations of purely technical discussions of human production and the importance of considering the competing interests that may be served by different production systems than are anthropologists. For example, Dourojeanni (1984), a forester, describes three land use capability studies in the same area of eastern Peru. The surveys' recommendations regarding the total land area to be left in forest cover ranged from five to sixty-five percent, while estimates of the land suitable for permanent cultivation ranged from eight to forty-two percent. All three studies were conducted by competent teams, and the figures presented by each are defensible on technical grounds. What distinguished them were contrasting po-

litical visions—of the institutions sponsoring the studies and the people they contracted to carry out the work—of what development in the region is supposed to accomplish, and for whom.

3. These paragraphs draw heavily on critical discussions of *The People of Puerto Rico* by Roseberry (1978) and Wolf (1978, 1990), which readers interested in a fuller treatment of the significance of the Puerto Rican study to anthropological method and theory should consult.

4. For a critique of this point of view, see Roseberry 1984.

REFERENCES

AN (Anthropology Newsletter)
1989 Conspicuously Absent. Anthropology Newsletter 30 (8): 2.

Bakx, K.
1987 Planning Agrarian Reform: Amazonian Settlement Projects, 1970–86. Development and Change 18 (4):533–55.

Benton, T.
1989 Marxism and Natural Limits: An Ecological Critique and Reconstruction. New Left Review 178:51–86.

Binswanger, H. P.
1991 Brazilian Policies that Discourage Deforestation in the Amazon. World Development 19 (7):821–29.

Brenner, R.
1977 The Origins of Capitalist Development: A Critique of Neo-Smithian Marxism. New Left Review 104:25–92.

Collins, J. L.
1986 Smallholder Settlement of Tropical South America: The Social Causes of Ecological Destruction. Human Organization 45 (1): 1–10.

Cook, S.
1973 Production, Ecology, and Economic Anthropology: Notes toward an Integrated Frame of Reference. Social Science Information 12 (1):25–52.
1974 Economic Anthropology: Problems in Theory, Method, and Analysis. *In* Handbook of Social and Cultural Anthropology, ed. J. J. Honigmann, 795–860. Chicago: Rand McNally.

Dourojeanni, M.
1984 Potencial y Uso de los Recursos Naturales: Consideraciones Metodológicas. *In* Población y Colonización en la alta Amazonía Peruana. Ed. Consejo Nacional de Población y Centro de Investigación y Promoción Amazónica, 110–21. Lima: Consejo Nacional de Población and Centro de Investigación y Promoción Amazónica.

Dunkerley, J.
1987 Power in the Isthmus: A Political History of Modern Central America. London: Verso Books.

Ehlers, T. B.
1990 Central America in the 1980s: Political Crisis and the Social Responsibility of Anthropologists. Latin American Research Review 25 (3):141–55.

Feit, H. A.
1986 Anthropologists and the State: The Relationship between Social Policy Advocacy and Academic Practice in the History of the Algonquian Hunting Territory Debate, 1910–50. Paper presented to the Annual Meeting of the American Anthropological Association, 3–7 December, Philadelphia.

Grundmann, R.
1991 The Ecological Challenge to Marxism. New Left Review 187:103–20.

Hardin, G.
1968 The Tragedy of the Commons. Science 162:1243–48.

Herskovits, M.
1951 Man and His Works. New York, N.Y.: Alfred A. Knopf.

Hjort, A.
1982 A Critique of "Ecological" Models of Pastoral Land Use. Nomadic Peoples 10:11–27.

Horowitz, M. M
1986 Ideology, Policy, and Praxis in Pastoral Livestock Development. *In* Anthropology and Rural Development in West Africa, ed. M. M Horowitz and T. M. Painter, 251–72. Boulder, Colo.: Westview Press.

Jones, J. C.
1991 Economics, Political Power, and Ethnic Conflict on a Changing Frontier: Notes from the Beni Department, Eastern Bolivia. IDA Working Paper 58. Binghamton, N.Y.: Institute for Development Anthropology.

Kroeber, A. L.
1939 Cultural and Natural Areas of Native North America. University of California Publications in American Archaeology and Ethnology 48:1–242. Berkeley: University of California Press.

Leff, E., ed.
1986 Los Problemas del Conocimiento y la Perspectiva Ambiental del Desarrollo. México City: Siglo Veintiuno Editores.

Little, P. D.
1985 Absentee Herders and Part-Time Pastoralists: The Political Economy of Resource Use in Northern Kenya. Human Ecology 13 (2):131–51.

Mintz, S. W.
1977 The So-Called World System: Local Initiative and Response. Dialectical Anthropology 2 (4):253–70.

Painter, M.
1990 Development and Conservation of Natural Resources in Latin America. *In* Social Change and Applied Anthropology: Essays in

Honor of David W. Brokensha, ed. M. S. Chaiken and A. K. Fleuret, 231–45. Boulder, Colo.: Westview Press.

Roseberry, W.
1978 Historical Materialism and The People of Puerto Rico. Revista/Review Interamericana 8 (1):26–36.
1984 Peasants, Commodities and Basic Foods. Paper presented to session "Capitalism and Its Consequences: Political Economic Studies of Economic Underdevelopment and Malnourishment in the World System Context," Annual Meeting of the American Anthropological Association, Denver, Colo.

Sandford, S.
1983 Management of Pastoral Development in the Third World. Chichester: John Wiley & Sons.

Schmidt, A.
1971 The Concept of Nature in Marx. London: New Left Books.

Schmink, M., and C. H. Wood
1987 The "Political Ecology" of Amazonia. *In* Lands at Risk in the Third World: Local-Level Perspectives, ed. P. D. Little and M. M Horowitz, 38–57. Boulder, Colo.: Westview Press.

Speck, F. G.
1915 The Family Hunting Band as the Basis of Algonkian Social Organization. American Anthropologist 17 (2):289–305.

Staff
1956 The Cultural Background of Contemporary Puerto Rico. *In* The People of Puerto Rico, by J. H. Steward, R. A. Manners, E. R. Wolf, E. Padilla Seda, S. W. Mintz, and R. L. Scheele, 29–89. Urbana: University of Illinois Press.

Steward, J. H.
1955 Theory of Culture Change: The Methodology of Multilinear Evolution. Urbana: University of Illinois Press.
1956 Introduction. *In* The People of Puerto Rico, by J. H. Steward, R. A. Manners, E. R. Wolf, E. Padilla Seda, S. W. Mintz, and R. L. Scheele, 1–27. Urbana: University of Illinois Press.

Vayda, A. P.
1967 On the Anthropological Study of Economics. Journal of Economic Issues 1:86–90.
1969 Comments on Dalton's Review Article. Current Anthropology 10:95.

Vessuri, H. M. C.
1986 Antropología y Ambiente. *In* Los Problemas del Conocimiento y la Perspectiva Ambiental del Desarrollo, ed. E. Leff, 203–22. México City: Siglo Veintiuno Editores.

White, L.
1949 The Science of Culture: A Study of Man and Civilization. New York, N.Y.: Grove Press.

Wissler, C.
1938 The American Indian: An Introduction to the Anthropology of the New World. 3d Ed. New York, N.Y.: Oxford University Press.

Wolf, E. R.
1972 Ownership and Political Ecology. Anthropological Quarterly 45 (3):201–5.
1978 Remarks on The People of Puerto Rico. Revista/Review Interamericana 8 (1):17–25.
1990 Distinguished Lecture: Facing Power—Old Insights, New Questions. American Anthropologist 92 (3):586–96.

Part 1
Central America

Chapter 1

Rethinking the Hamburger Thesis: Deforestation and the Crisis of Central America's Beef Exports

Marc Edelman

Since the early 1970s, growing numbers of tropical ecologists, anthropologists, and other social scientists have linked developed-country demand for Latin American beef to a host of environmental and social ills, including forest destruction (Buschbacher 1986; DeWalt 1983; Heckadon and McKay 1984; Nations and Kromer 1983a, 1983b; Parsons 1976; Shane 1986; Partridge 1984; Uhl and Parker 1986a), decreased rainfall (Hagenauer 1980; Fleming 1986), soil erosion (Myers 1981; Nations and Nigh 1978), replacement of food crops by pasture (Boyer 1986; Feder 1980; DeWalt 1983; Sanderson 1986; Spielman 1972), reduced developing-country protein consumption (Buxedas 1977; Caufield 1985; Dickinson 1973; Holden 1981; Roux 1975), rural unemployment (Brockett 1988; Williams 1986), and concentration of land, credit, and other resources (Aguilar and Solís 1988; Da Veiga 1975; Edelman 1985; Guess 1978; Keene 1978; Rutsch 1980; Slutsky 1979).[1] Articles have appeared with great frequency in normally staid scientific publications, with uncharacteristically catchy titles referring to the "hamburger connection" (Myers 1981), "hamburger society" (Nations and Kromer 1983a), "cattle eating the forest" (DeWalt 1983), and "our steak in the jungle" (Uhl and Parker 1986a).[2] One recent book on the Central American environmental crisis labels the region's countries "hamburger republics" (Hedström 1985: 41). Headlines of journal commentaries inquire rhetorically "Is a quarter-pound hamburger worth a half-ton of rainforest?" (Uhl and Parker 1986b) and "Is the rainforest worth seven hundred million hamburgers?" (Matteucci 1988).

This equation of export beef production with environmental destruction has also taken on a life of its own in the popular imagination and in the mainstream communications media, in part because of newly urgent concerns about global warming, declining species diversity, and an eroding land base (Durning 1991). Advocacy groups, conjuring up images of cattle stampeding through fragile rainforests (EPOCA n.d.), have launched boycotts of fast-food chains, staged sit-ins at the World Bank to protest the negative environmental impact of lending policies, and urged major cutbacks in herd sizes and beef consumption (RAN 1989; Beyond Beef Coalition, The Goal: A 50% Reduction of Beef Consumption by 2002, *The New York Times,* 23 April 1992: B7). Editorial writers who for years proclaimed the advantages of the Brazilian "economic miracle" now decry "the rape of Rondônia" and rail against the "murder" of Alaska's temperate rain forest, proclaiming that "Tongass Trees Aren't Cheeseburgers" (Forest Murder: Ours and Theirs, *New York Times,* 20 September 1989: A26).[3]

The widespread fetishization of the hamburger-deforestation relation could be considered a successful diffusion of scientists' concerns to a broader public. But although not inaccurate *grosso modo*, like most fetishes it both reveals and conceals. As a focus for anxieties about the global environment, it is appealingly direct and suggests possibilities for effective personal action that are attractive in the individualistic and depoliticized yet health-conscious United States of the late 1980s and 1990s (boycotting Burger King, for example). As a framework for social scientific analysis, however, the by-now familiar allusions to ground-beef-for-gringos emerging from tropical pastures sometimes obscure more than they elucidate. Furthermore, oversimplifications about the "hamburger connection" may give rise to erroneous environmental policy recommendations and political strategies.

This chapter argues that the demand-based understandings of the beef export-deforestation relation that underlie the prevailing wisdom—popular and/or scientific—are no longer valid for Central America in the period of livestock sector stagnation that dates to the early 1980s. Even when applied to the deforestation that occurred during the beef export boom of the 1960s and 1970s, such explanations frequently require significant qualification, because they ignore or downplay both other factors that fueled the beef boom and other causes of forest destruction. Cattle ranching has been associated with deforestation; this chapter suggests that the relationship is a historically specific one and is, in any case, more complex than has sometimes been appreciated.[4] Adequate policy prescriptions must be based on an understanding of that complexity (Allen and Barnes 1985; Rudel 1989; WRI 1985).

Central America has been the United States' third most important source of imported beef, after Australia and New Zealand, since the mid-1950s.[5] This chapter examines: (1) the diminishing role of foreign demand as a stimulus to livestock production in Central America, the principal source of U.S. tropical beef imports; (2) the causes of the crisis in the Central American livestock sector, especially in Costa Rica, the region's main exporting country and one much admired for its innovative conservation programs (Tangley 1986; Thrupp 1989); and (3) the extent to which the crisis in the Central American livestock sector creates political and physical space for environmentally and economically sound alternatives to pasture development and forest destruction.

Foreign Demand and Social Scientific Manichaeanism

Social scientists concerned with change in the developing world are only beginning to recover from the Manichaean excesses of dependency and world-system theory that were so fashionable in the 1960s and 1970s.[6] In their worst manifestations, these demand-based or circulationist paradigms posited an all-powerful metropolitan "capitalism" as the explanation for underdevelopment in the periphery and in effect denied that "local initiative and local response" (Mintz 1977) had any significant role in making history. Now that historical process and human agency again occupy their deserved, privileged place in social scientific investigation (Ortner 1984), it is perhaps surprising that dualist, dependency-type determinisms remain dominant in certain areas of inquiry, such as in discussions of the beef exports-deforestation relationship.

It is ironic that rising public concern about export-oriented ranching and tropical forest destruction coincides not only with diminishing interest in dependency-type paradigms in the social sciences—an esoteric consideration understandably of no concern to environmental policymakers and activists—but more importantly also with significant declines in the foreign demand for beef that is supposed to be the root of the problem. Indeed, in Central America, a key beef-producing zone that sends most of its exports to the United States, the cattle sector has been in a serious crisis since the early 1980s (see figs. 1 and 2). In Costa Rica, the severity of the situation is such that the Federation of Chambers of Cattle Ranchers (*Federación de Cámaras de Ganaderos*) successfully petitioned in mid-1985 to have the Minister of Agriculture declare a

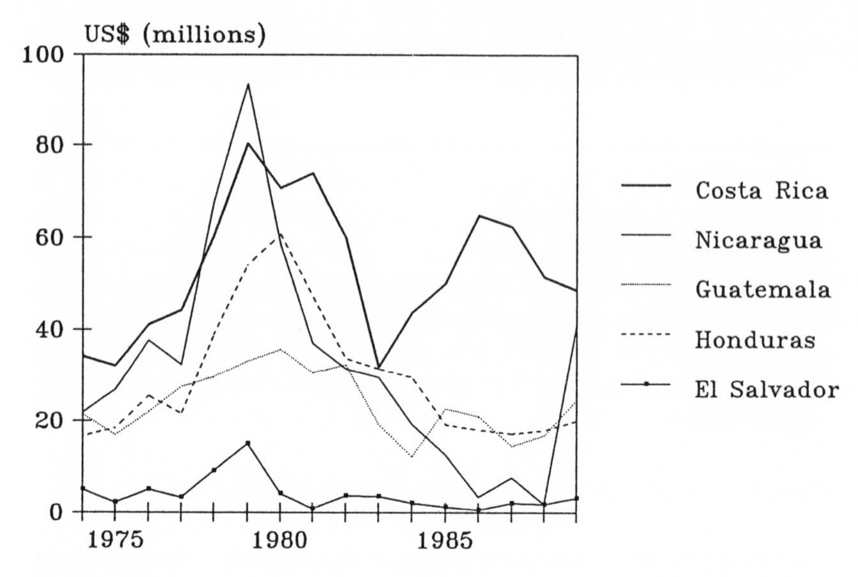

Fig. 1. Central America: value of beef exports, 1974–89. (Data from FAO, Trade Yearbooks.)

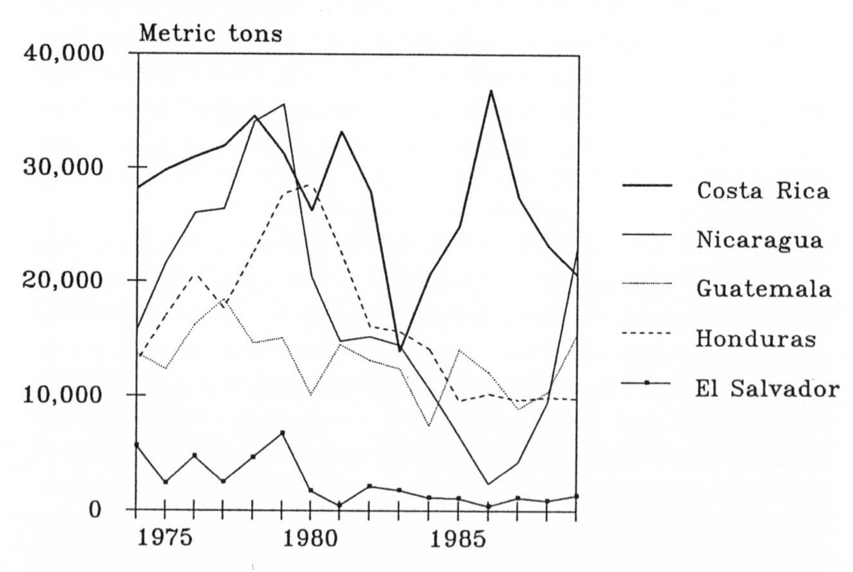

Fig. 2. Central America: volume of beef exports, 1974–89. (Data from FAO, Trade Yearbooks.)

"state of emergency" for the livestock sector. Central Bank experts concurred, noting that for cattle production "the rate of return was negative" (Realidad 1988:8). "When there's no meat," the president of the Cattle Ranchers Federation warned in 1988, "we'll see the politicians coming to the producers, trying to correct their errors" (Montenegro 1988:9).

This rancher pessimism, even if hyperbolic at times, hardly conforms to the popular image of voracious, avaricious cattlemen bulldozing ever-greater expanses of virgin forest. The livestock "emergency" is real, although its origins are diverse and are only partially attributable to contraction of U.S. demand. Few recent analyses of the "hamburger connection" recognize the crisis (Annis 1990 is one exception), and some that do erroneously attribute it solely to the 1979 Nicaraguan revolution, when pro-Somoza ranchers drove their herds to neighboring Costa Rica and Honduras (Nations and Leonard 1986: 73).[7] In fact, the crisis has both foreign—that is, U.S.—and domestic Central American roots. Both point to the need for political-economic analyses of how forces of supply and demand are mediated and shaped by the U.S. market and political system and by domestic Central American institutions, interest groups, and economic actors.

Given the intense anxiety aroused among Central American ranchers and foreign trade ministers, it is surprising that hardly any recent discussions of the "hamburger connection" mention the U.S. curbs on beef imports that went into effect at the end of 1979. Technically a "countercyclical" amendment to the 1964 Meat Import Act (P.L. 88-482), this measure (P.L. 96-177) provided for decreasing import quotas during expansive phases of the U.S. cattle cycle and increasing them during periods of contraction.[8] P.L. 96-177 has meant that the U.S. market is unlikely to absorb any significant increases in Central America's exportable beef surpluses.[9] Domestic U.S. livestock producers, concerned about low-priced beef imports, fought long and hard for passage of P.L. 96-177. They were opposed by lobbyists for the fast-food industry and consumer groups that feared higher beef prices, as well as by representatives of Central American and Caribbean beef-exporting nations. The details of this battle are beyond the scope of this discussion, but it should be sufficient to indicate that, at the very least, the strength of demand itself, rather than being determined either by capitalism or by North American hamburger hunger, is inseparable from processes of political struggle between contending interest groups. And—also rarely mentioned in the "hamburger" literature—per capita North American beef consumption has also been

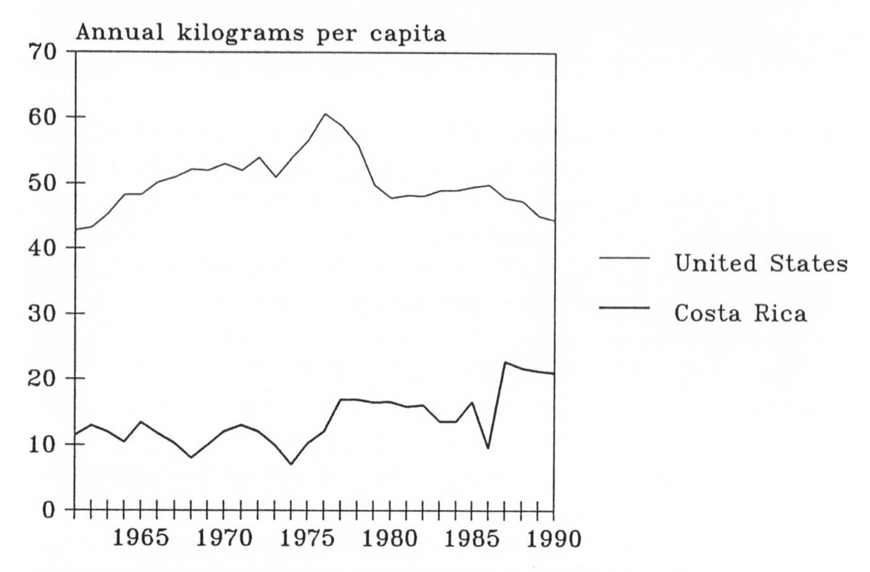

Fig. 3. Costa Rican and U.S. beef consumption, 1961–90. (Data from U.S. Department of Agriculture.)

dropping since 1976 (see fig. 3), even though the proportion derived from cheap grass-fed beef, like that exported from Central America, has been on the rise.

The crisis in the Costa Rican cattle sector in the 1980s is not only the result of rigid U.S. demand codified by the 1979 countercyclical legislation. Ranchers who sold fattened cattle to export packinghouses traditionally had paid rather little in taxes: a one percent tax on the value of each steer, eleven colones in various municipal levies, five colones for a stamp to finance the Rural Guard, and similarly symbolic assessments based on the weight of the on-hoof animal and the meat extracted.[10] In December 1981, following the previously stable colón's plunge from 8.6 to 38 per dollar in little more than a year, the Legislative Assembly, responding to importers' calls for relief, established a tax intended to limit exporters' windfall profits. In the case of the cattle sector, the law provided for a ten percent tax on "the difference between the exchange rate at which hard currency from export earnings is sold and the total that the same sale would have produced had it been realized at 8.60 colones per U.S. dollar" (CLD 1981, Laws 6707 and 6696). Suddenly, in 1981–83, cattlemen who historically had scarcely considered export taxes (or other taxes for that matter—see Edelman 1992:251) a cost of doing business were facing payments of 500 to 1,000 colones (approximately $12.50 to $25.00 in early 1982) for each steer sold to the packing plants. While this represented only about five percent of the value of an average steer, it cut dangerously into the narrow 6.5 percent average profit margin for beef producers that had been established in agreements with the export packers (FCG 1983:16). By the mid-1980s, some ranchers claimed that the total tax burden for each exported steer had risen to over 30 percent of the animal's value (Realidad 1988).

Just as ranchers had little experience with taxation, they were similarly unaccustomed to operating in genuine financial markets, in spite of the growing disaffection of many in the 1970s and 1980s with the costs of Costa Rica's social democratic development model and a consequent infatuation with pro-laissez-faire rhetoric. The unprecedented inflation of the early 1980s brought a steep rise in interest rates that represented a major additional burden to producers with low profit margins and few sources of working capital. Costa Rican ranchers had long taken for granted that the National Banking System would provide credit at negative real rates of interest, allowing them a hedge against inflation or funds that could be diverted, albeit without authorization, to more lucrative activities. While the high inflation of 1981–82 initially produced a sharp drop in real interest rates, by 1983–84 most loan rates had recovered to the point where ranchers, now also beset by soaring costs for veterinary and other inputs, registered a mounting number of defaults. By early 1987, nearly two-thirds of Costa Rican banks' cattle loans were in arrears (Annis 1990).

International beef prices, measured in constant dollars, have also declined precipitously since the early 1970s, hardly an incentive for expanding herds or carving more pastures out of the forest, even if the costs of land clearing and grass planting are absorbed by peasant colonists or tenants. Indeed, as figure 4 indicates, by the 1980s beef prices were at or below the levels prevailing during the first years of the beef boom in the 1950s and early 1960s.

Deforestation Trends

Data on forested and deforested area must be taken with some caution, since the methods and definitions employed by different agencies and in different time series may not be consistent (Allen and Barnes 1985). Figure 5 provides three estimates of deforestation trends in Costa Rica,

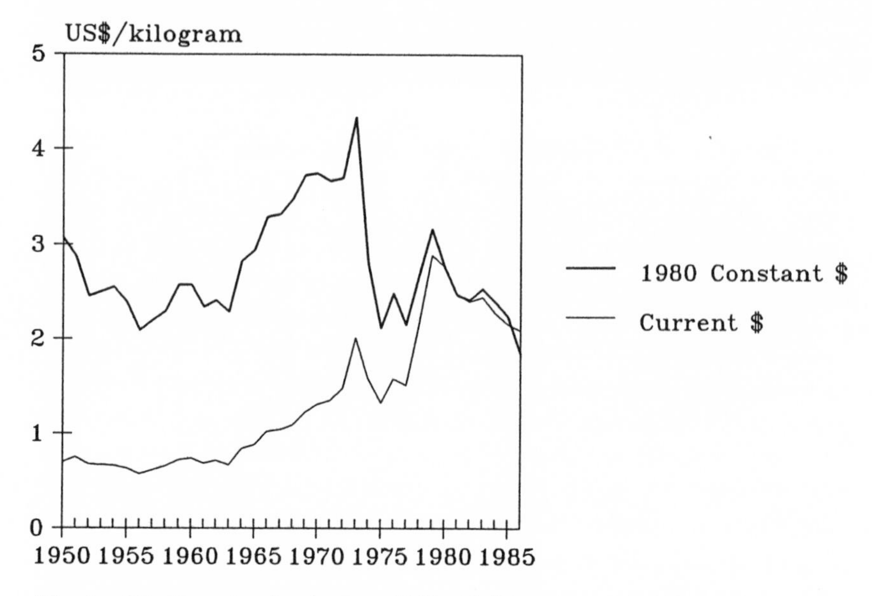

Fig. 4. World market beef prices, 1950–86. (Data from World Bank Commodity Trade and Price Trends, 1987.)

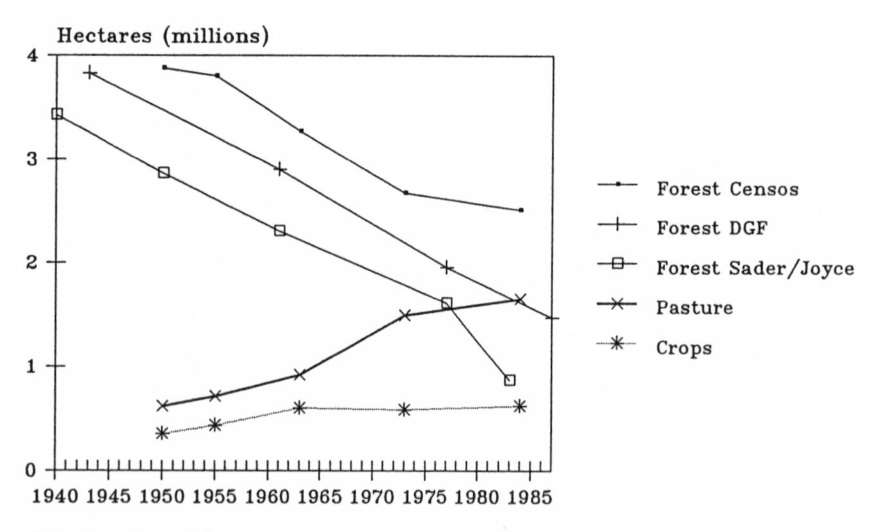

Fig. 5. Costa Rican land use, 1940–87. (Data from *Dirección General Forestal* and *Censos Agropecuarios* [pasture and crops].)

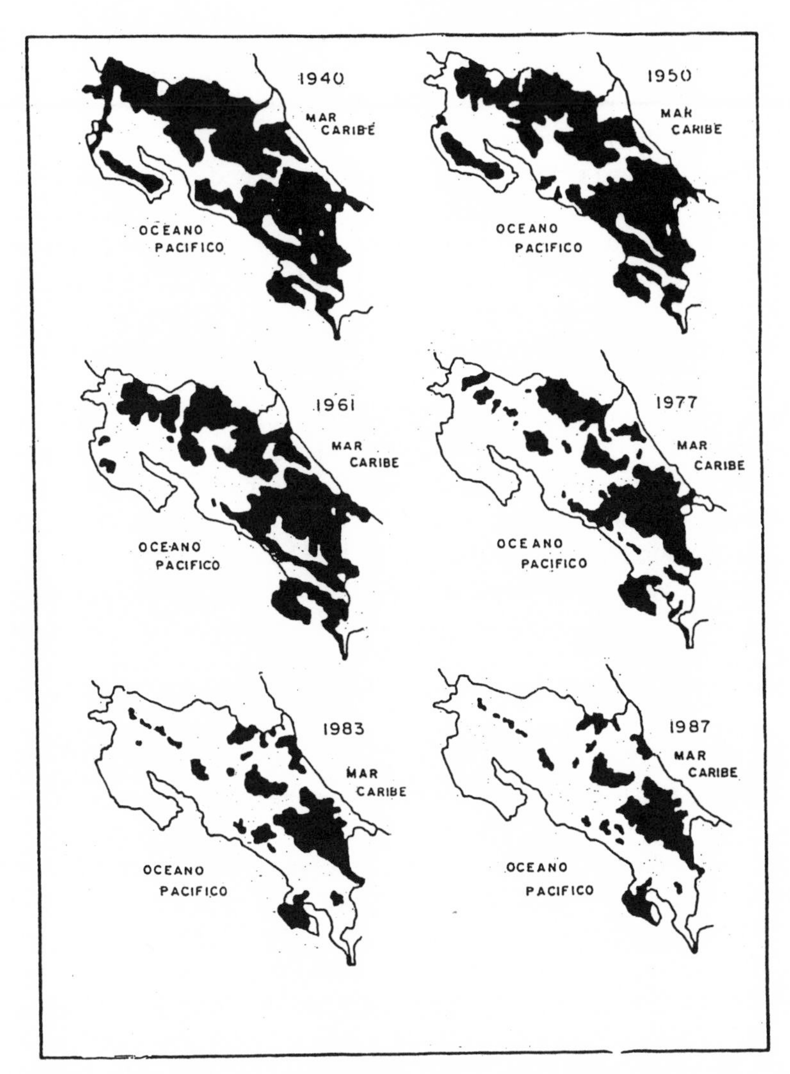

Fig. 6. Costa Rica: area of closed canopy forest. (Data from *Dirección Forestal.*)

as well as data on pasture expansion and cultivated area. The destruction of Costa Rica's forests over the past five decades is alarming by any measure (see fig. 6).[11] Nevertheless, the "least pessimistic" estimate, derived from the last five agricultural censuses, suggests that rates of deforestation and pasture growth have slowed in the latest period.[12] This estimate would suggest that pasture expansion approached (and indeed exceeded) natural limits (in the 1963–73 intercensal period) and then slowed with the recent lack of dynamism in the livestock sector (in the last years of the 1973–84 period).

The other two deforestation estimates are based on differing interpretations of maps derived from similar sets of aerial photographs and satellite (Landsat) data; they employ a stricter definition of "forest" and a broader one of "forest clearing."[13] Sader and Joyce's estimate for the 1977–83 interval, the only departure from otherwise remarkably linear long-term trends, bears special examination both because it is extremely alarming and because it received prominence in a recent highly publicized report (WRI, UNEP, and UNDP 1990). These authors suggested that tiny Costa Rica lost an average of 124,000 hectares of primary forest each year during the seven-year period. But they also conceded that their editing of the 1983 data "may have resulted in an overestimation of forest clearing in the [1977–83] period" (1988:17), a caveat that passed unnoticed by both the authors of the WRI, UNEP, and UNDP report and media alarmists who accorded Costa Rica the dubious honor of "first place for deforestation in Latin America" (La Nación 1990a). The findings that caused such shock in 1990 were based on Sader and Joyce's estimate for 1977–83 (WRI, UNEP, and UNDP 1990:102–3, 292–93).

This estimate is approximately twice as high as the most extreme on-the-ground measures and was very likely (as the authors' proviso indicates) influenced by map editing procedures. Ground-based measures are hardly free of problems, but they do suggest some slowing of deforestation in recent years. During 1950–83, deforestation averaged 53,382 hectares per year (this still represents deforestation of approximately one percent of the national territory and three percent of remaining forested area each year, an extremely high rate compared to other developing countries);[14] in 1984, a total of 41,875 hectares were cleared (16,875 of which were authorized by the state); by 1985, estimates of deforestation by different government agencies ranged from 13,000 to 29,250 hectares (of which 10,750 hectares were authorized) (SEPSA 1986a, 1986b). In 1990, Raúl Solórzano, president of Costa Rica's Tropical Science Center, recognized that "the country is now in a transition stage. Wood [not cattle] is becoming the most important factor in deforestation" (David Dudenhoefer, Forest Crisis Nears, *Tico Times,* 23 February 1990:10).[15]

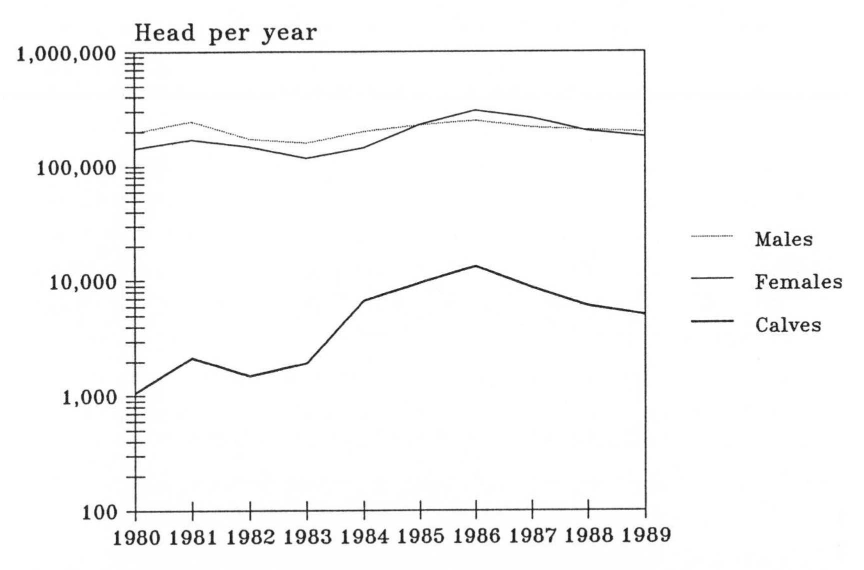

Fig. 7. Costa Rican cattle slaughter by age/sex, 1980–89 (semilogarithmic scale). (Data from *Consejo Nacional de Producción.*)

Domestic Supply and Demand

In Costa Rica, the crisis in the cattle sector has been so pronounced that during much of the 1980s ranchers slaughtered the herd faster than it could reproduce.[16] Sluggish export demand, higher taxes and indebtedness, and soaring interest rates and veterinary input costs have made cattle ranching a losing proposition for most investors (more on this below). The only ranchers able to endure are those who do not require bank financing, who obtained land and herds at little or no cost through inheritance, who specialize in breeding expensive exhibition bulls, or who in effect subsidize cattle operations with profits from other production lines. These tend to be the largest haciendas, rather than the small- and medium-sized producers. Figure 7 illustrates the dramatic increase since the early 1980s in the slaughter of cows and calves (note that on a semilogarithmic graph, lines with similar slopes indicate similar proportional rather than similar absolute changes). This is indicative of ranchers' interest in liquidating existing investments, but it is distinguished from their traditional reactions to cyclical downturns in beef markets by a new indifference about maintaining breeding stock for future expansion.[17]

The notion that export beef production was a scheme foisted on naive

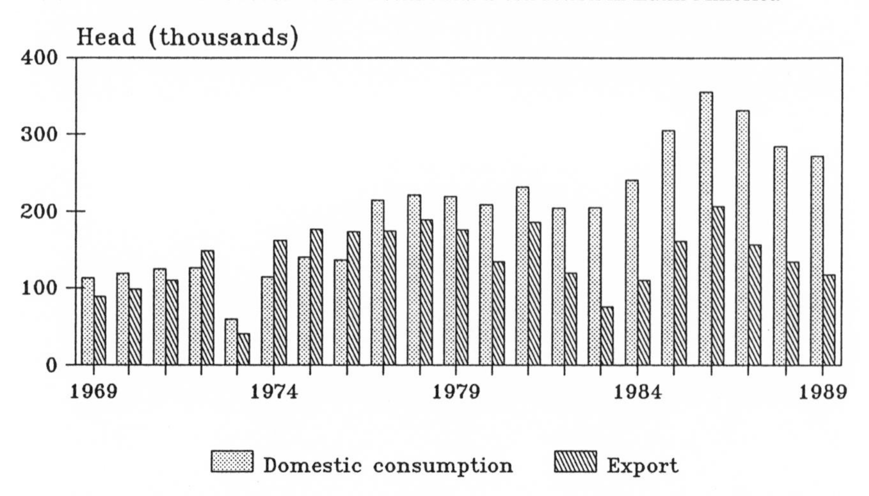

Fig. 8. Costa Rican beef production, domestic consumption, and exports, 1969–89. 1973 data is for six months. (Data from *Consejo Nacional de Producción.*)

and vulnerable developing countries by international lending institutions (occasionally, one hears, in cahoots with U.S. fast-food chains) is only partially correct. Developing-country governments have been understandably reluctant to refuse offers of credit, but the recurrent dependency-type interpretations of this phenomenon ignore the extent to which cattle interests in each producing country organized, pressured, and adopted new technologies, all with the objective of entering international markets. Once this was achieved—in the 1950s for most of the region, earlier for Nicaragua, later for El Salvador—ranchers' lobbies have been among the most active political pressure groups, working to widen access to foreign markets and to secure favorable pricing, credit, and land tenure policies from regional governments. In Costa Rica, for example, leading ranchers articulated a vision of their country as a beef exporter à la Texas or Argentina even in the 1930s, when domestic needs still had to be satisfied with large imports from Nicaragua and more than two decades before significant shipments were sent abroad.[18] In the post-1955 period, the cattle lobby has been a constant presence in national politics (Edelman 1992:chap. 10; Annis 1990). Its more prominent members have also significantly diversified their interests in other economic sectors, including export agriculture, agroindustry, retail commerce, tourism, and communications media (Aguilar and Solís 1988).

While it is true that the advent of the beef export economy in Central America was associated everywhere with plummeting per capita beef

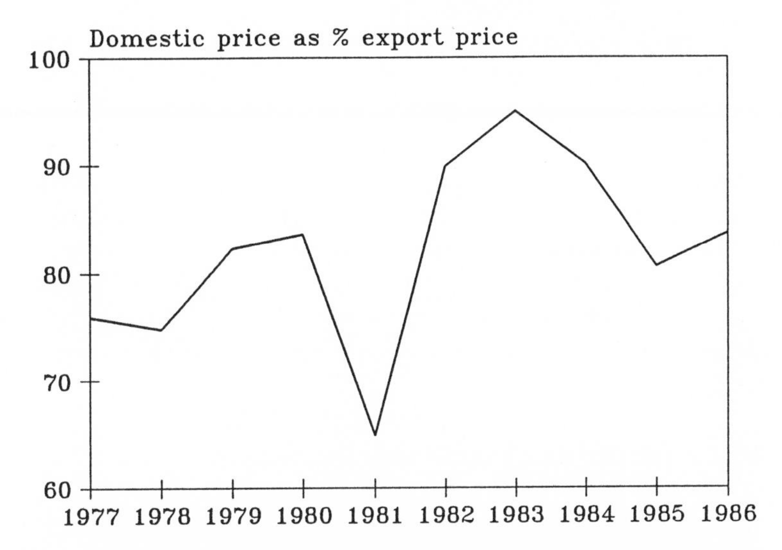

Fig. 9. Costa Rican export and domestic beef prices, 1977–86. (Data from *Banco Central de Costa Rica,* Cifras sobre producción agropecuaria.)

consumption, the expansion of herds to meet foreign demand eventually reversed this decline in most of the region (Williams 1986).[19] Until the crisis of the mid-1980s, when even heifers began to be liquidated, only steers were sent to the export slaughterhouses. But to breed more steers, ranchers had to acquire increasing numbers of cows, and it was these that supplied much of the domestic demand and brought per capita consumption back up to near earlier, preexport boom levels (see fig. 3). In Costa Rica, with sluggish exports and stagnant international prices, domestic demand is increasingly important as a proportion of the total slaughter (see fig. 8).[20] This reflects both herd liquidation, taking place for reasons noted above, and a convergence since the late 1970s between domestic and world beef prices that reduces tendencies to channel as many head as possible to export packinghouses (see fig. 9).[21]

Domestic demand has received insufficient attention in the "hamburger connection" literature, something that is particularly unfortunate in a period when it is of much greater relative weight as a stimulus to cattle production in Central America, Brazil, and elsewhere. It has long been recognized that meat is a product with a very high income elasticity of demand; in other words, small increases in income levels produce proportionally greater increases in demand. In the absence in many countries of per capita real income increases during the economic crisis of the 1980s, population growth and urbanization are still producing rapidly rising demand for beef. Growth rates for domestic demand now exceed those for production even in many traditional exporting countries, such as Brazil, Mexico, El Salvador, Guatemala, Nicaragua, and Panama (Jarvis 1986). Normally, such a situation might be expected to produce rising prices and profits. To understand why such conditions have not been sufficient to rescue the livestock sector from crisis, it is necessary to consider how cattle enterprises differ from other types of production and why ranchers do not always appear to behave as rational economic actors operating in a perfect market with various types of investment options.

Subsidies from Nature, Subsidies from the State

Tropical ranching has always depended on a heavy subsidy from nature, whether the availability of low-cost land, the one-time extraction of nonlivestock forest resources such as timber, or the "production" of semiferal cattle on vast estates (Bunker 1985; Edelman 1985, 1992). In mixed crop-livestock enterprises, cattle have traditionally served as insurance against higher risks inherent in cash crop cultivation. Increasingly, the economic rationality of ranching has hinged as well on significant state subventions, including favorable fiscal, land titling, and road-building policies; artificially low interest rates; price controls and duty exonerations for inputs; and government technical assistance programs (Annis 1990; Binswanger 1989; Browder 1988; Hecht et al. 1988). Usually, it has been based on extracting these diverse income streams—ground rent, speculative rent, and institutional rent (de Janvry 1981; Edelman 1985)—rather than on high profitability according to accepted accounting conventions. The illegal diversion of subsidized cattle production credit to more profitable sectors of the economy and the use of ranches principally as collateral for bank loans are only two of the most common scams that proliferate in tropical America. In some countries—notably Brazil—planting grass on untitled state land is considered a demonstration of the effective occupation required to establish secure claims and eventual private property rights; elsewhere, such "improvements" on the land protect owners from expropriation under agrarian reform laws, especially if a few cattle are grazing the newly planted pasture.

Latin American governments have rarely established significant legal sanctions against deforestation. The 1990 Forest Law in Costa Rica provides maximum prison terms of three years for those engaging in illegal forest felling on public land and one year for those doing so on private property. It also creates financial mechanisms that assure the continued existence of an enforcement and oversight apparatus (La Gaceta 1990). Most of the region's legal codes, however, do not even define deforestation, and resources for enforcement are nearly nonexistent (Porras and Villarreal 1986).[22]

The debt crisis and inflation of the 1980s have affected the complex system of subsidies for livestock in a variety of ways. In many countries, the most direct institutional rent streams—artificially low interest rates, price supports, and various kinds of tax privileges—have been cut significantly with the imposition of new neoliberal models of economic development. Even where—as in Costa Rica—concerted struggle by cattle interests has succeeded in maintaining some access to below-inflation-rate loans, this frequently has not been sufficient to cure past defaults or to offset soaring input costs and land prices that augment the opportunity cost—in nonfrontier areas—of using expensive properties for grazing.[23] Throughout the continent, high indebtedness has reduced states' capacities for providing direct subventions to uneconomic productive sectors, even if they are politically influential.[24] Especially when the foreign exchange generated is not commensurate with resources invested or land area occupied, austerity-conscious policymakers have been less and less inclined to submit to the demands of cattle lobbies and their allies. And the leading ranchers, long adept at converting rent income into capital, have diversified to the point where their coherence as a distinct, unified interest group can no longer be taken for granted. Indeed, many have embraced "eco-tourism" and new nontraditional exports (cf. Bulmer-Thomas 1988), and all the associated incentives, with the same ardor with which they earlier pursued low-cost cattle loans.

Environmental Conservation in a Period of Crisis

The crisis in the Central American livestock sector could, under certain conditions, be a blessing in environmental terms. If a changed economic climate is no longer impelling ranchers to destroy forest, political and economic space could conceivably be opened for the discussion and implementation of sustainable development and conservation measures.

Several difficulties exist, however. First and most obvious is the extreme lack of resources for funding sustainable development and conservation programs (a problem not unrelated, as indicated below, to stagnation of the beef cattle sector itself). Second is the assumption—explicit or implicit in much of the "hamburger" literature—that forests are destroyed to produce cattle per se, rather than to generate income whatever the source. Third, commonly proposed solutions, such as debt-for-nature swaps or labor-intensive, sustainable exploitation of remaining forest resources, are likely to be of relatively limited scope, especially—in the latter case—in Central America. And finally, reforestation efforts, spurred by generous fiscal incentives, have sometimes tended to reinforce the skewed resource and income distributions that were characteristic of the cattle economy and that contributed indirectly to environmental degradation in the first place.

Livestock specialists have long known that it is technically feasible to produce considerably more beef per hectare with relatively small additional investments in pasture management, breeding, veterinary inputs, and human care and administration (León et al. 1981; Leonard 1985; Parsons 1976). Indeed, in Costa Rica, holdings of less than 20 hectares already produce an average of 383 kilograms of beef per hectare per year, while those over 200 hectares produce 162 kilograms (Hartshorn, Hartshorn, Atmella et al. 1982:6). In seasonally dry tropical environments, such as Central America's Pacific coastal plain, the better pastures can—even with investments like those presently made by many small ranchers—sustain greater grazing densities than the one-head-per-hectare level that has remained virtually unchanged since at least 1950 (León et al. 1981).

Yet even during the beef boom the latifundist land tenure system in most cattle zones has encouraged extensive land use, as ranchers seeking to extract natural capital in the form of grass- or brush-fed beef expanded into unused portions of their estates or surrounding land. Many of these landowning families acquired properties decades or even centuries ago and are thus under little pressure to modernize production or to consider the opportunity cost of land in their profit-loss calculations (Edelman 1985, 1992). The structure of incentives that has shaped modern cattle production does little or nothing to distinguish between high quality and marginal pasture. It fosters neither grazing intensification on existing pastures in areas where this would be feasible nor the abandonment and reclamation of suboptimal grasslands. In the absence of such stimuli, the landscape of Central America's depressed cattle zones will continue to have the haunting, semiabandoned quality that has become increasingly typical in the late 1980s, with mixed brush and pastures stretching as far as the eye can see, and only scattered small animals that will be slaughtered before reaching full finished weight.

Restructuring incentives would require not only the political will to confront what is, in Costa Rica at least, a still resilient cattle lobby, but financial resources that are simply no longer available in an era of economic adjustment and retrenchment. Ironically, demand-based explanations of the "hamburger connection" and their "logical" correlates, demand-reducing policy prescriptions, have an unintended and potentially damaging role in exacerbating this resource scarcity. In a trenchant denunciation of hamburgerist reasoning and activism, Daniel Janzen, whose conservationist credentials hardly brook criticism,[25] argues that North American consumers should buy, not boycott, Costa Rican beef (Janzen 1988). Refusing to buy beef (or any other tropical product) from Central America, Janzen reasons, will simply lead to more of it being purchased from other tropical zones, accelerating deforestation there. Just as important, the damage to developing countries' foreign trade balances reduces "the tax and resource base that is picking up the bill for the national parks and other kinds of reserves" (Janzen 1988:258).[26] While Janzen's argument is perhaps most compelling as regards Costa Rica, the nation with the strongest commitment to environmental conservation (and the least acute social crisis), all other Central American beef exporters, including even tiny El Salvador, have taken significant steps towards establishing protected natural areas.

Though much of the "hamburger" literature notes that pasture development is usually the last and most drastic of a series of exploitative uses of forest (or former forest) land, discussion has often taken on an almost teleological tone, assuming that beef rather than money is the ultimate goal of those destroying the tropics. Were more attention devoted to logging, artisanal and industrial gold mining, peasant squatting, and plantation agriculture, it would perhaps become clearer that the natural capital of the forest can be used to subsidize a variety of activities. In the absence of strict enforcement of forest reserve boundaries and, even more importantly, of economic opportunities for those displaced by, among other things, extensive grazing and capital-intensive agriculture and industry, the pressure on remaining natural capital is not likely to abate. Moreover, the relative intensity of distinct pressures on forest resources can and does vary considerably even within a small geographic area, such as Central America. Rather different policy tools are likely to be required for addressing the destruction caused by logging in Guatemala's Petén, gold mining in Costa Rica's Osa Peninsula, or smallholder squatting on the Nicaraguan or Honduran agricultural frontiers. The success of such policies needs to be appraised not only by the purely "physical/biological criteria" (Thrupp 1989) on which conservationists typically rely but also by socioeconomic measures that consider the well-being of rural populations, particularly in the areas of year-round income and employment generation and assuring access to land for cultivation.

Extractive Reserves

Recent efforts (Allegretti 1990; Clay 1989; Peters et al. 1989) to demonstrate that sustained yield gathering of nonwood forest products in the Brazilian and Peruvian Amazon can produce economic benefits per unit of area greater than swidden agriculture or ranching have excited considerable interest among environmentalists (and among ice cream enthusiasts, with the marketing by Vermont manufacturer Ben and Jerry's of a new line of products flavored with exotic jungle fruits and nuts). But in spite of promising dessert recipes and the best of intentions, this labor-intensive, forest-sensitive strategy would probably confront some formidable obstacles if implemented on a large scale, especially in Central America. Unlike the Amazonian Indians, *caboclos* and *ribereños* that have long included petty commerce in gathered forest products (nuts, fruits, resins, etc.) as one among several complementary survival strategies, Central America's forest or near-forest peoples are—with few exceptions—thoroughly "peasantized" in terms of their productive experiences and traditions, knowledge, and expectations.[27] Reconstructing and diffusing the forest lore necessary to make large-scale commercial gathering economically viable for marginal peasant families in Central America would require resources and time, neither of which are widely available, as well as demonstrating to policymakers and to skeptical and unskilled potential participants the feasibility of prompt returns. Cooperative management of forest resources, well developed in some regions of the Amazon and a necessary concomitant of such schemes, is basically unknown in Central America outside of the few remaining indigenous forest peoples. Few cultural models exist for resolving tensions between individual accumulation and the common good in favor of the latter.[28]

Commercial gathering of nonwood resources as a means of conservation has some significant inherent limits. Much depends on the physical characteristics of the gathered product, on who controls the initial stages of commercialization and processing, on the strength of demand, and on the availability or lack of alternative sources of supply. Perishability or essential processing by monopsonistic, non-gatherer-controlled agroindustries may affect negatively the prices paid to collectors. This in turn may lead some, in the absence of strong social controls, to compensate for low unit prices by extracting greater, perhaps unsustainable, volumes.[29] If significant social controls are present, they may be adequate to prevent overharvesting, but not to confront unfavorable market conditions or asymmetrical power relations between those at different points on the chain of commercialization.[30]

To have a genuine environmental impact, gathering and extractive reserve formation must be carried out on a large scale.[31] But unless demand expands as fast as supply, high output volumes are likely to bring falling prices.[32] This may tempt collectors once again to harvest the forest's capital individually rather than just the "interest" cooperatively. If demand and prices do increase in a sustained fashion, the emergence of alternative sources of supply—perhaps based on laboratory fabrication or plantation production of previously gathered products—cannot be discounted.[33] Precedents for such a scenario are notorious, such as pharmaceutical giant Eli Lilly's synthesis of a drug for Hodgkin's disease from the rosy periwinkle, a native plant of Madagascar, or the dramatic, early twentieth-century expansion of Malayan rubber cultivation with seeds exported from Brazil.

Successful gathering projects, then, must ideally be based on nonperishable, preferably uncultivatable products that can be harvested on a sustainable basis and that have reliable markets yet few alternative sources of supply. Extraction projects must involve substantial gatherer control over initial processing and marketing. They must also be carried out by peoples with traditions of communal resource management that in effect control against unsustainable individual exploitation of the environment. Regrettably, the very specificity of these conditions does not augur well for the generalization of an extractive reserve model on a scale that would permit it to have a major environmental effect.

Debt-for-Nature Swaps

Other mechanisms for stemming forest destruction that are now much in vogue, such as debt-for-nature swaps, are also, unfortunately, feasible primarily on a small scale. Such deals are, ironically, the bogey of the security-obsessed Brazilian right and the Central American left, both of which are suspicious of the implications for sovereignty.[34] In spite of having the appearance of "free" debt reduction, Latin American finance ministers, too, tend to regard these and other debt-equity trades with caution, fearing the inflationary consequences of the large injections of local currencies that widespread buy-back deals would entail (Patterson 1990:10). Such effects may be mitigated by issuing bonds in local currency with long amortization periods instead of simply printing more money (Sevilla 1990:148–154). But debtor governments using locally borrowed funds to pay foreign purchasers of their debt paper may ultimately face greater budget deficits and financial costs, since developing country interest rates tend to be higher than international levels in order to discourage capital flight (Hedström 1990:188).

These formidable political obstacles obviously do not preclude occasional successes, such as the swaps that have contributed to strengthening Costa Rica's exemplary system of parks and reserves. But the very infrequency of such experiences—and their concentration in a country with an unusual political culture, a tiny territory, and a unique history of indulgence by foreign creditors and aid agencies—suggests that swap schemes, while more than a drop in the bucket, are very far from being a panacea for the problems of either deforestation or debt.

Reforestation

Increasingly, in the semiabandoned savannas of Costa Rica's devastated cattle zones, one sees large, well-fenced plantations of uniform pochote (*Bambacopsis quinatun*) saplings, a high-priced hardwood, and occasionally other species such as melina (*Gmelina arborea*), madero negro (*Gliricidia sepium*), and teak (*Tectona grandis*). The reforested area is still pathetically small in relation to what is needed, however, although the rate of growth is high (Ugalde and Gregersen 1987). In 1979–1984, official data indicate that a total of 8,865 hectares were reforested, 5,562 hectares with fiscal incentives that permitted landowners an income tax deduction of approximately U.S. $1,000 per reforested hectare (SEPSA

1986b). In as little as ten years, the owners of the pochote farms in particular, after benefiting from generous tax breaks and low-interest loans, stand to reap a veritable bonanza, marketing the timber in central Costa Rica or abroad.

Between 1979 and 1988, the economic benefits of reforestation incentives accrued primarily to medium- and large-size enterprises, in part because the lowest income groups simply did not pay income tax and therefore did not seek such deductions. Moreover, larger investors, unlike most smallholders, tended to have sufficient collateral for bank loans and were able to survive the lean years between planting and harvesting (Porras and Villarreal 1986; Brenes 1988). The large farmer bias in reforestation policy often had counterproductive environmental consequences, in addition to problematical implications for social equity. "Once the procedure was established," a report sponsored by the U.S. Agency for International Development notes, "an interesting phenomenon occurred:

> the 51 applications covering an area of 11,234 hectares were presented by businessmen who were looking for ways to lower their taxes. The reforestation companies that appeared overnight started businesses whereby they offered to lower taxes for persons by selling them one or more reforested hectares and charging them for technical assistance, administration and the value of the land—almost like a real estate business. These projects did generate employment for workers, technicians and professionals as well as salesmen but the reforestation value is dubious. The projects are generally centrally located in areas suited to livestock and agriculture and have begun to compete with food production. In some cases, natural forests were cut to be later replanted under the program. Of the total area (11,234 hectares) included in the reforestation program up to 1981, 1,638 hectares or only 14 percent has been reforested. (Hartshorn, Hartshorn, Atmella et al. 1982:34)[35]

Since 1988, the policies that facilitated this type of corruption—what Costa Ricans term "sausages" (*chorizos*)—have been significantly modified, and smallholders have increasingly joined reforestation efforts. The major change was the creation of a "forestry bond" (*Certificado de Abono Forestal*), a negotiable instrument provided to reforesters (and worth, in 1991, approximately U.S. $1,000 for each hectare planted in trees, about half of initial production costs). Financed initially by the Dutch government through a U.S. $5 million debt swap with the Costa Rican government, this program was added to the existing program of tax write-offs, but it differed from the earlier system in that it provided "bonds" not only to well-off investors but to organized groups (e.g., cooperatives, peasant unions, cantonal agricultural centers) composed of individuals wishing to plant seedlings on up to 15 hectares each.[36] In order to maintain a rotating credit fund for continuing reforestation, beneficiaries are required to repay the bond after twenty years, approximately the time that they harvest the trees; they receive a seven-year grace period and pay an annual interest rate of only 8 percent, well below prevailing inflation levels (Sevilla 1990:151). In 1990, this system accounted for approximately 8,000 reforested hectares, 3,340 of which (42 percent) were planted by small producers' groups. Approximately 15,000 hectares are needed to meet Costa Rica's annual consumption of wood products (Se reforestará 15 mil hectáreas en 1991, La Nación, 2 October 1991).

Official statistics, however, may understate the actual extent of reforestation, just as they may obscure the variety of motives impelling smallholders to plant trees. Even with the advent of forestry bonds, current policy continues to encourage the planting of dense single-species stands and official data reflect this preference. Peasant tree farming of mixed stands of native or exotic species (for wind barriers, fuel, posts, shade, nuts, and fruit, as well as for lumber) represents a significant environmental factor, albeit one that is difficult to administer or measure precisely (Jones and Price 1985). Small-scale tree farmers, by conserving high biotic diversity, also reduce the risk, inherent in official reforestation schemes, of host-specific pests and pathogens devastating single-species stands.

Heavily indebted smallholders—and there are many, given the crisis in Costa Rica's agrarian sector—have also found that reforestation (or preservation of existing woods) is one of the few ways of emerging from default and regaining access to production credit. Increasingly, peasants whose properties are subject to bank liens and potential foreclosures have negotiated arrangements in which they use trees as collateral for old debts or new loans. Sometimes a smallholder divides a property and consolidates debts on a section with remaining timber, allowing the rest of the farm or the trees themselves to serve as a guarantee for a new crop production loan. In other cases, part of the farm may be reforested, with future lumber harvests functioning as security.[37] These innovative initiatives, without parallel elsewhere, may well increase in the future, as smallholders become better acquainted with the practice and as banks develop relevant procedures.[38]

One need not be overly sanguine about the ultimate environmental effect of smallholder involvement in agroforestry, whether tree plantation style or indigenous style, to recognize that in a relatively brief period Costa Rica has taken some significant strides. Before the establishment of smallholder reforestation programs, conservationists had to trade off a potentially salutary environmental impact against significant social costs. Because they incorporated previously excluded sectors of the rural population into reforestation efforts, they have reforested a larger area each year, and at least some of the negative distributional and employment effects of the 1979–88 programs have been mitigated. Importantly, such effects, like those of the beef cattle boom, were among the factors that contributed to building pressure on Costa Rica's resource base in the first place.

Conclusion

This chapter has examined the "hamburger-deforestation" literature in light of the crisis in Central America's (and especially Costa Rica's) beef exports during the 1980s. It argues that both social scientific analyses and the popular awareness to which they have contributed have lagged behind current realities in failing to take note of the inadequacies of demand-based explanations of the beef export-deforestation relation. Indeed, the "hamburger-deforestation" literature is among the last bastions of a simplistic version of dependency theory that ignores the extent to which political contention influences the strength of U.S. demand for foreign beef, the important role that Central American political actors have had in shaping the conditions in which cattle producers operate, and the complexity of the deforestation process itself.

The severe crisis in the Costa Rican cattle sector in the 1980s resulted not only from lower international prices and politically influenced contraction in U.S. demand, but also from new fiscal, pricing, and credit policies that to a large degree reflected outcomes of domestic Costa Rican political processes. The crisis in beef exports coincided with strengthened domestic demand, an increasingly important factor in many beef-producing countries that has received no attention in the "hamburger" literature. But domestic demand did not usually generate the revenue traditionally produced from exports; increased domestic consumption was essentially opportunistic, deriving instead from the liquidation of herds by bankrupt, disillusioned ranchers. Whatever the causes of the crisis, however, a stagnant cattle export sector, even if buoyed by opportunistic domestic demand, is unlikely to be a major force behind further forest destruction. Much of Costa Rica's pasture land has, in fact, been abandoned in recent years, suggesting that if deforestation continues (as it has, though at a reduced rate), other types of explanations and solutions must be sought.

This requires rethinking not only the fundamental premises of the "hamburger" literature, but also those of some of the more popular policy remedies now frequently raised in discussions of environmental degradation in Central America. These include demand-reducing beef boycotts, extractive reserve models derived from different historical and cultural contexts, reforestation incentive programs for large farmers, and the debt-for-nature swaps that, as indicated above, are only feasible on a very small scale that permits little more than a highly localized environmental impact. The beef export economy was fueled by a massive subvention from nature and wasteful transfers of public resources to a relatively modest number of large ranchers who long enjoyed favorable fiscal policies, as well as subsidized credit, inputs, and technical assistance, and privileged access to the centers of political power. This skewing of income and wealth distributions, and the accompanying distortions of otherwise democratic political processes, led a wide range of actors—from rich ranchers to poor, displaced squatters—to reproduce themselves on the basis of a natural capital that lasted approximately one generation. Some of these actors had (or could have found) viable alternatives; others did not have other options. The challenge for this and future generations is to find environmentally sound measures that permit reclaiming a damaged land base, while ensuring that equity and social welfare are not sacrificed in the process.

NOTES

1. Since 1989 when I presented a draft of this chapter to the American Anthropological Association in Washington, several authors have advanced similar arguments about deforestation and cattle (e.g., Harrison 1991; Lehmann 1991, 1992) or about extractive reserves (Browder 1992). I am reassured, as any

researcher must be, that in considering similar data they arrived at basically parallel conclusions. I wish to express my appreciation to Claudia Alderman, William Durham, Ricardo Godoy, Angelique Haugerud, Jeffrey R. Jones, Michael Painter, Juan Pablo Ruiz, and Robert G. Williams for their constructive criticisms of the manuscript and to Jayne Hutchcroft and César Rodríguez for assisting with the research. Obviously, I alone am responsible for any factual or analytical errors.

2. For the author's own modest contribution to this genre, see Edelman 1987.

3. Alarm about cattle and deforestation has also been utilized to demonize national enemies. In hearings on Brazil's role in Iraq's military buildup, Senator Albert Gore claimed erroneously that "the biggest single customer for beef coming out of the Amazon is—has been Iraq" (U.S. Congress, Transcript of Hearings of the Subcommittee on Science and Technology of the Joint Economic Committee, 21 September 1990, cited in Sanderson and Capistrano 1991:39).

4. Harrison (1991:90), analyzing Costa Rican agricultural census data from 1950–84, found a very strong correlation at the cantonal level between forest loss and pasture expansion. It is unlikely, for reasons outlined in this chapter, that this association would be very strong for the latter half of the 1973–84 intercensal period or subsequent years. In the absence of true time-series data or a more recent census, however, this is a difficult relation to test.

5. South American countries have not been permitted to export fresh or chilled beef to the U.S. market because of the presence of *aftosa* (hoof-and-mouth disease). Traditionally, they had their major markets in Europe, but EEC protectionism has increasingly limited this source of foreign earnings. Separated from South America by the Darien Gap forests, *aftosa*-free Central America, Mexico, and the Caribbean are the only suitable sources of imported fresh beef in the hemisphere, according to the U.S. Department of Agriculture.

EEC subsidies have, in recent years, permitted the continent to enter the ranks of exporters, and even Brazil has occasionally imported large quantities of European beef (LACR 1987), suggesting that foreign beef demand is no longer a major force behind the continuing destruction of the Amazon. Certainly U.S. demand is irrelevant in this respect: Browder (1988:115) points out that the Brazilian Legal Amazon supplied 1,700 metric tons of beef to the United States in 1982, a typical recent year, accounting for a whopping seven one-thousandths of one percent (.00007) of U.S. consumption. Clearly, at the very least, the widespread notion that North American meat lust is fueling the destruction of the Amazon cannot be sustained. Brazil, however, had a very rapidly growing domestic demand for beef until the 1988 economic collapse (LACR 1988).

6. Representative works are Frank (1969) and Wallerstein (1974). Much ink has been spilled in polemics about this approach, based in many cases on textual exegeses of Marx's writings on capitalism. Mintz (1977) provides a concise, historically based critique that, unlike much of the literature, does not descend into either undue abstraction, quasi-scriptural appeals to the authority of Marxist texts, or turgid prose.

7. One might ask why Costa Rica's and Honduras's beef exports did not rise with this new influx of *somocista* cattle. In Costa Rica, at least, the herd declined significantly by the mid-1980s, making the significance of Nicaraguan imports as an explanation for dropping exports all the more problematical. The decline in Nicaragua's exports was due to particular circumstances: liquidation of herds by pro-Somoza ranchers in 1979 and the early 1980s, a reflection of political fears and artificially low prices; and the 1985 U.S. trade embargo. One year after the Reagan administration's imposition of trade sanctions, the U.S. Department of Agriculture proposed removing Nicaragua from the list of countries eligible to export meat to the United States, ostensibly because the "dangerous situation" there prevented inspection of local ranches (LACR 1986b). Political instability in El Salvador also had a devastating effect on the cattle sector, with the herd declining from 1.2 million head in 1979 to 600,000 in 1983, the lowest level in fifty years (CAR 1984).

8. On the U.S. cattle cycle's effects in Central America, see Edelman (1987:543–44). For more detailed analyses, see Jarvis (1986:55–64) and Sanderson (1986:156–64).

9. An additional U.S.-imposed obstacle is the institution in 1983 of tougher quality control standards for meat imports. These had a particularly devastating impact on Honduras, the least-developed Central American economy (LACR 1986a), though several other producing nations were affected as well.

10. These were used to finance the Livestock Department of the National Production Council (*Consejo Nacional de Producción*—CNP), the Animal Health Section of the Ministry of Agriculture, the *Federación de Cámaras de Ganaderos* and the regional *Cámaras*, and the Veterinary Medicine Faculty of the National University. An additional 1 percent was withheld to guarantee payment of income tax.

11. It is worth noting that most export beef production and associated deforestation has been centered in traditional livestock zones along the seasonally dry Central American Pacific coastal plain and—in Nicaragua—in the Chontales region east of Lake Nicaragua. Technically, this did not involve rainforest destruction, the bugaboo of much current discussion, but rather the felling of dry or semihumid tropical forest and the replacement of natural (or ancient derived) savanna with African grasses. If there is to be sustainable development of a grass-fed cattle sector, it is most likely to occur in these zones (Daubenmire 1972a, 1972b, 1972c). Pasture expansion in rainforest areas (e.g., Guatemala's Petén, eastern Nicaragua, and eastern Costa Rica) has sometimes occurred to supply domestic needs as traditional areas shifted to export production. Often, though, ranching expansion into humid forest zones is the second or third stage of more complex processes of colonization that have involved squatter activity, road building, logging, and plantation agriculture (Carrière 1990; Repetto 1990).

12. This estimate was derived by adding the total forested land in private holdings to the total of untitled state lands (which were assumed to be forested). Obviously, this has limitations. Much forested land on farms is likely to be

somewhat degraded or intervened in; similarly some untitled land undoubtedly consists of secondary growth or degraded primary forest.

13. Sader and Joyce (1988:123) define "forest clearing" as removal of 20 percent or more of the upper canopy. The criteria employed in the DGF estimates are not specified with such precision; the DGF states that "forested area" refers to "closed canopy forest."

14. Estimates of the area in closed canopy forest in the late 1970s vary from 34 to 41 percent of the national territory of 50,900 square kilometers (Porras and Villarreal 1986: 20–21). Sader and Joyce's 1983 satellite imagery analysis claimed that only 17 percent of the country was covered with primary forest (Sader and Joyce 1988). For deforestation rates in other third world countries, see Allen and Barnes (1985) and WRI, UNEP, and UNDP (1990).

15. Similarly, in Honduras, a 1992 government report on deforestation found that livestock raising was not an important factor. The three most significant reasons for forest destruction were cutting for firewood (the source of 70 percent of the country's energy consumption), forest fires caused by swidden cultivators, and industrial logging (La integridad del bosque y la política forestal, *El Tiempo,* August 1992:6). Utting (1991) provides an insightful discussion of the many causes of deforestation in different countries in the region.

16. The CNP, the state commodities agency, estimated in 1985 that for the herd to maintain its size, no more than 30.7 percent of male animals, 10.5 percent of cows, and 0.7 percent of calves ought to be slaughtered. The actual figures were 36.1 percent of male animals, 16.9 percent of females, and 1.8 percent of calves. The female and male animals weighed, on average, 6.4 and 8.6 percent less, respectively, than those slaughtered the previous year (SEPSA 1986a:16).

17. In 1982, it was estimated that, of Costa Rica's bovine population of 2.3 million, 46.2 percent consisted of beef cows, 9.8 percent "double-purpose" (dairy and beef) cows, and 11.2 percent dairy cows. Males (beef and dairy) accounted for only 32.8 percent of the total. Almost all of these were beef steers; only 1.8 percent of the total population were stud animals and 0.7 percent oxen (SEPSA 1983).

18. This is discussed in more detail in Edelman 1992: chap. 5.

19. This has not prevented some analysts (e.g., Caufield 1985; Hedström 1985) from implying that this trend continued long after it in fact had been reversed.

20. In figure 8, data before 1974 refer to twelve-month "cattle years" from July to June of the following year; 1973 data are for the second half of 1973. After 1973, years are calendar years.

21. With the exception of 1981, when major devaluations occurred that brought exporters extraordinarily high colón prices. Without this outlier, 54 percent of the 1977–86 variation in the domestic price/international price ratio is explained by the linear trend.

22. "Deforestation" is, as Hamilton remarks, a "horribly ambiguous word" that has referred to "fuelwood cutting; commercial logging; shifting cultivation;

fodder lopping; forest clearing for conversion to continuous annual cropping, to grazing, to tree crops (foods, extractives, beverages), to forest plantations; and burned or grazed areas that are still essentially in forest" (1991:5).

23. In 1988, for example, the Ministry of Agriculture and Livestock (MAG) agreed to subsidize for three years the cost of cattle loans granted by the Banco Nacional, the principal lender. Rates for small, medium, and large producers were set, respectively, at 9, 12, and 18 percent—in a year when the overall inflation rate was 25 percent (MAG y BNCR Formalizan Subsidio para Ganaderos, *La Nación,* 25 August 1988). Ultimately, the difference almost certainly must be financed by foreign borrowing (or by the Central Bank printing more currency).

24. Where—as in Brazil—governments have long mouthed support for market economics (even while implementing Keynesianism for the rich), the rationality of pasture expansion appears to be tied most closely to the need to establish possession rights, rather than to continuing flows of subsidized credit, which have been substantially reduced in recent years. Nevertheless, the tax advantages that come with possession and the hedge it represents against inflation are still important (Binswanger 1989; Fearnside 1989; Hecht et al. 1988).

25. Janzen has been a prime mover behind the establishment of the Guanacaste National Park (GNP) in northwestern Costa Rica, an innovative conservation area, much of which is on former cattle land (Janzen 1986; McLarney 1988). Even so, the project has been assailed as an "appropriation of a huge area of land by a privileged North American who, even though good-hearted and well-intentioned, is perpetuating inequitable patterns of development and privatization [which] . . . conflict with the interests of the land-hungry poor" (Thrupp 1989:8).

Thrupp's contention that GNP represents a "private acquisition" is a variant of the suspicion many Latin Americans feel about foreign conservationists and their goals. In the case of the GNP, this likely derives from the role of the Nature Conservancy and similar organizations in raising funds for purchasing land and for an endowment to cover operating costs. Both title to park land and control of the endowment will belong to the Costa Rican National Park Foundation (*Fundación de Parques Nacionales* or FPN). Critics have charged, however, that the nongovernmental FPN constitutes a "parallel" structure that undermines sovereignty, since it duplicates functions of the state-run National Park service (*Servicio de Parques Nacionales* or SPN) but does not enjoy the FPN's access to foreign financial resources (Vargas 1990: 14). For a sophisticated, balanced discussion of foreign conservationists' involvement in Latin America, see Sevilla (1990).

26. Jagels (1990) argues in a similar vein that consumer boycotts of tropical forest hardwoods are counterproductive.

27. The exceptions, often members of indigenous groups or corporate peasant communities with common land, include *xate* palm harvesters, allspice gatherers, and chicle tappers in Guatemala's Petén (Nations 1992) and resin

tappers in Honduras (Stanley 1991). Even the rural Amazonian groups that rely on extraction as one part of their survival strategy also engage in more destructive activities (gold panning, livestock raising, swidden agriculture, hunting, overharvesting of extracted resources) and are only arguably defenders of tropical forests (Browder 1992).

28. Harvester inexperience and laziness can also be obstacles to sustainable management of forest resources (Nations 1992:216). For a rich discussion of the complexities of managing communal resources, see McCay and Acheson (1987).

29. In Costa Rica, the extraction of edible "hearts" (*palmito*) from the *pejibaye* (*Guilielma utilis*) and similar palms led to their virtual disappearance in some zones. Not surprisingly, given high foreign and domestic demand, this once-wild species is now widely cultivated.

30. Few studies of extractive reserves or "tolerant forest management" practices have benefited from studies of markets distant from project zones, even though export of gathered products is often a key goal. This introduces an element of significant uncertainty into project planning, in addition to weakening the antideforestation arguments of reserve advocates.

31. The environmental impact of extractive activities has, however, received little attention. Pearce (1990:47) suggests, for example, that harvesting Brazil nuts "can have a devastating effect on the birds and animals that would otherwise have eaten them."

32. After the creation of widely publicized extractive reserves in the Brazilian state of Acre, "prices for natural rubber and Brazil nuts have dropped sharply, jeopardizing the people's livelihood" (IDB 1992). Part of the rubber price drop was due to the removal of tariffs that kept the domestic price of rubber at roughly three times the world level. This raises the question of whether any renewable tropical forest resource can provide a basis for sustained economic activity without external subsidization (Browder 1992:176). Brazil nut collectors, even in 1990, received only 2 to 3 percent of the New York wholesale price of their nuts (Pearce 1990:1046). This share may rise to as much as 10 percent if producers are able to bypass intermediaries and transport the nuts to the local processing factory (Clay 1992:403–4), though it is uncertain if such steps will be sufficient to offset price declines.

33. Brazil nuts, for example, are now a wild forest product, the seeds of a tree (*Bertholletia excelsa*) that are sometimes propagated on a small scale by farsighted gatherers. The success of such informal propagation efforts and the creation of relatively dense single-species stands (by Amazonian standards, at least) suggest that the tree might be suitable for plantation cultivation if economic conditions warrant.

34. This lack of symmetry in the way different groups on the political spectrum have reacted to the debt-for-nature concept in Brazil (where it is supported by important elements of the left) and in Central America and Bolivia (where its backers tend to be rightist or centrist) obviously cries out for further analysis. Sevilla (1990) argues that in the Bolivian, Costa Rican, and Ecuadorian cases, governments and local conservation organizations have been important partici-

pants in negotiations (and later in administration of protected areas) and that the fears regarding sovereignty are groundless.

35. Much of the remaining 9,596 hectares (86 percent) likely consisted of land on which planted seedlings did not survive. See Thrupp (1989) and Annis (1990) for similar descriptions of the distortions produced by reforestation tax incentives.

36. Holland acquired U.S. $33 million of Costa Rican debt from U.S. banks for U.S. $5 million. In return for retiring this obligation, the Costa Rican Central Bank supplied U.S. $10 million in colones for use in reforestation programs.

37. Interviews in Guácimo de Limón, July 1990.

38. Costa Rica's 1990 Forest Law established that standing trees may guarantee loans from public sector banks, but it left to the banks the job of drawing up pertinent regulations (La Gaceta 1990:5).

REFERENCES

Aguilar, Irene, and Manuel Solís
1988 La Élite Ganadera en Costa Rica. San José: Editorial de la Universidad de Costa Rica.

Allegretti, Mary Helena
1990 Extractive Reserves: An Alternative for Reconciling Development and Environmental Conservation in Amazonia. *In* Alternatives to Deforestation: Steps toward Sustainable Use of the Amazon Rain Forest, ed. Anthony B. Anderson, 252–64. New York: Columbia University Press.

Allen, Julia C., and Douglas F. Barnes
1985 The Causes of Deforestation in Developing Countries. Annals of the Association of American Geographers 75 (2):163–84.

Annis, Sheldon
1990 Debt and Wrong-Way Resource Flows in Costa Rica. Ethics and International Affairs 4:1–15.

BCCR (Banco Central de Costa Rica)
1988 Cifras sobre Producción Agropecuaria. San José: BCCR.

Binswanger, Hans P.
1989 Brazilian Policies that Encourage Deforestation in the Amazon. Environment Department working paper no. 16. Washington, D.C.: World Bank.

Boyer, Jefferson C.
1986 Capitalism, Campesinos and Calories in Southern Honduras. Urban Anthropology 15 (1–2):3–24.

Brenes Castillo, Carlos
1988 ¿Desarrollo Forestal Campesino? *In* La Situación Ambiental en Centroamérica y el Caribe, ed. Ingemar Hedström, 163–74. San José: Departamento Ecuménico de Investigaciones.

Brockett, Charles D.
1988 Land, Power, and Poverty: Agrarian Transformation and Political Conflict in Central America. London: Unwin Hyman.
Browder, John O.
1988 The Social Costs of Rain Forest Destruction: A Critique and Economic Analysis of the "Hamburger Debate." Interciencia 13 (3):115–120.
1992 The Limits of Extractivism: Tropical Forest Strategies beyond Extractive Reserves. BioScience 42 (3):174–182.
Bulmer-Thomas, Victor
1988 The New Model of Development in Costa Rica. *In* Central America: Crisis and Possibilities, ed. Rigoberto García, 177–196. Stockholm: Institute of Latin American Studies.
Bunker, Stephen G.
1985 Underdeveloping the Amazon. Urbana: University of Illinois Press.
Buschbacher, Robert J.
1986 Tropical Deforestation and Pasture Development. BioScience 36 (1):22–28.
Buxedas, Martín
1977 El Comercio Internacional de Carne Vacuna y las Exportaciones de los Países Atrasados. Comercio Exterior (Mexico) 27 (2):1494–1509.
Carrière, Jean
1990 The Political Economy of Land Degradation in Costa Rica. New Political Science 18/19:147–163.
Caufield, Catherine
1985 In the Rainforest. New York: Knopf.
CAR (Central America Report [Guatemala])
1984 Beef Exports Earn Fewer and Fewer Dollars. Central America Report 11 (34) (31 August):270.
Clay, Jason
1992 Buying in the Forests: A New Program to Market Sustainably Collected Tropical Forest Products Protects Forests and Forest Residents. *In* Conservation of Neotropical Forests: Working from Traditional Resource Use, ed. Kent H. Redford and Christine Padoch, 400–415. New York: Columbia University Press.
Clay, Jason W., ed.
1989 Indigenous Peoples and Tropical Forests. Cambridge, Mass.: Cultural Survival.
CLD (Colección de Leyes y Decretos)
1981 Colección de Leyes y Decretos 1981. San José: Imprenta Nacional.
Daubenmire, R.
1972a Some Ecological Consequences of Converting Forest to Savanna in Northwestern Costa Rica. Tropical Ecology 13 (1):31–51.
1972b Standing Crops and Primary Production in Savanna Derived from

Semideciduous Forest in Northwest Costa Rica. Botanical Gazette 133 (4):395–401.
1972c Ecology of Hyparrhenia Rufa in Derived Savanna in Northwestern Costa Rica. Journal of Applied Ecology 9:11–23.

Da Veiga, José S.
1975 A la Poursuite du Profit: Quand les Multinationales Font du "Ranching." Le Monde Diplomatique September:12–13.

de Janvry, Alain
1981 The Agrarian Question and Reformism in Latin America. Baltimore: Johns Hopkins University Press.

DeWalt, Billie R.
1982 The Big Macro Connection: Population, Grain and Cattle in Southern Honduras. Culture and Agriculture 14:1–12.
1983 The Cattle Are Eating the Forest. Bulletin of the Atomic Scientists 39 (1):18–23.

Dickinson, Joshua C.
1973 Protein Flight from Latin America: Some Social and Ecological Considerations. *In* Latin American Development Issues, Proceedings of the Conference of Latin Americanist Geographers, ed. David Hill, 3:127–132. East Lansing, Mich.: CLAG Publications.

Durning, Alan B.
1991 Fat of the Land. World Watch 4 (3):11–17.

Edelman, Marc
1985 Extensive Land Use and the Logic of the Latifundio: A Case Study in Guanacaste Province, Costa Rica. Human Ecology 13 (2):153–185.
1987 From Central American Pasture to North American hamburger. *In* Food and Evolution: Toward a Theory of Human Food Habits, ed. Marvin Harris and Eric B. Ross, 541–561. Philadelphia: Temple University Press.
1992 The Logic of the Latifundio: The Large Estates of Northwestern Costa Rica since the Late Nineteenth Century. Stanford, Calif.: Stanford University Press.

EPOCA (Environmental Project on Central America)
n.d. Central America: Roots of Environmental Destruction. San Francisco: EPOCA.

FCG (Federación de Cámaras de Ganaderos de Costa Rica)
1983 Problemática de la Ganadería Bovina y Propuestas para su Reactivación en el Corto Plazo. San José: FCG.

Fearnside, Philip M.
1989 A Prescription for Slowing Deforestation in Amazonia. Environment 31 (4):16–20, 39–40.

Feder, Ernest
1980 The Odious Competition between Man and Animal over Agricultural Resources in the Underdeveloped Countries. Review 3 (3):463–500.

Fleming, Theodore H.
1986 Secular Changes in Costa Rican Rainfall: Correlation with Elevation. Journal of Tropical Ecology 2:87–91.
Frank, André Gunder
1969 Capitalism and Underdevelopment in Latin America. New York: Monthly Review Press.
La Gaceta
1990 Ley No. 7174: Reforma a La Ley Forestal. La Gaceta Diario Oficial 112 (133) (16 July):1–7.
Guardia Quirós, Jorge
1987 La Política de Precios en Costa Rica. San José: COUNSEL.
Guess, George
1978 Narrowing the Base of Costa Rican Democracy. Development and Change 9 (4):599–609.
Hagenauer, Werner
1980 Análisis Agró-Metereológico en la Zona de Cañas y Bagaces (Guanacaste) en los Años 1921 a 1979. Informe Semestral [Instituto Geográfico Nacional], July–December:45–59.
Hamilton, Lawrence S.
1991 Tropical Forest Misinterpretations. Journal of Forestry 89 (1):5–6.
Harrison, Susan
1991 Population Growth, Land Use and Deforestation in Costa Rica, 1950–84. Interciencia 16 (2):83–93.
Hartshorn, Gary, Lynne Hartshorn, Agustín Atmella, et al.
1982 Costa Rica Country Environmental Profile. A Field Study. San José: Tropical Science Center—U.S. Agency for International Development.
Hecht, Susanna B., Richard B. Norgaard, and Giorgio Possio
1988 The Economics of Cattle Ranching in Eastern Amazonia. Interciencia 13 (5):233–240.
Heckadon Moreno, Stanley, and Alberto McKay, eds.
1984 Colonización y Destrucción de Bosques en Panamá. Panama: Asociación Panameña de Antropología.
Hedström, Ingemar
1985 Somos Parte de un Gran Equilibrio. La Crisis Ecológica en Centroamérica. San José: Departamento Ecuménico de Investigaciones.
1990 ¿Volverán las Golondrinas? La Reintegración de la Creación desde una Perspectiva Latinoamericana. San José: Departamento Ecuménico de Investigaciones.
Holden, Robert H.
1981 Central America is Growing More Beef and Eating Less, as the Hamburger Connection Widens. Multinational Monitor 2 (10): 17–18.
IDB (Inter-American Development Bank)
1992 Brazil: New Programs for Environment. The IDB 19 (5):11.

Jagels, Richard

1990 Soothing the Conscience: Tropical Forest Exploitation Revisited. Journal of Forestry 88 (10):27–31.

Janzen, Daniel H.

1986 Guanacaste National Park: Tropical Ecological and Cultural Restoration. San José: Editorial Universidad Estatal a Distancia.

1988 Buy Costa Rican Beef. Oikos 51:257–258.

Jarvis, Lovell S.

1986 Livestock Development in Latin America. Washington, D.C.: World Bank.

Jones, Jeffrey R., and Norman Price

1985 Agroforestry: An Application of the Farming Systems Approach to Forestry. Human Organization 4 (4):322–31.

Keene, Beverly

1978 La Agroindustria de la Carne en Costa Rica. San José: Confederación Universitaria Centroamericana. Mimeo.

LACR (Latin American Commodities Report [London])

1986a Meat / Honduras. Latin American Commodities Report CR-86-08 (25 April): 8.

1986b Meat / Nicaragua. Latin American Commodities Report CR-86-13 (10 July): 8.

1987 Little Cheer for Beef Markets, says FAO. Latin American Commodities Report CR-87-04 (26 February):4–5.

1988 Brazil Cattlemen Oppose Import Request. Latin American Commodities Report CR-88-14 (15 August):14.

Lehmann, Mary Pamela

1991 After the Jungleburger: Forces behind Costa Rica's Continued Forest Conversion. Latinamericanist (Center for Latin American Studies, University of Florida) 26 (2):10–16.

1992 Deforestation and Changing Land-Use Patterns in Costa Rica. *In* Changing Tropical Forests: Historical Perspectives on Today's Challenges in Central and South America, ed. Harold K. Steen and Richard P. Tucker, 58–76. Durham, N.C.: Forest History Society and IUFRO Forest History Group.

León, Jorge S, Carlos Barboza V., and Justo Aguilar

1981 Desarrollo Tecnológico en la Ganadería de Carne. San José: Consejo Nacional de Investigaciones Científicas y Tecnológicas. Mimeo.

Leonard, H. J.

1985 Natural Resources and Economic Development in Central America: A Regional Environmental Profile. Washington, D.C.: International Institute for Environment and Development.

Matteucci, Silvia Diana

1988 Is the Rain Forest Worth Seven Hundred Million Hamburgers? Interciencia 12 (1):5.

McCay, Bonnie J., and James M. Acheson, eds.
1987 The Question of the Commons: The Culture and Ecology of Communal Resources. Tucson: University of Arizona Press.
McLarney, William O.
1988 Guanacaste: The Dawn of a Park. Nature Conservancy Magazine 38 (1):11–15.
Mintz, Sidney
1977 The So-Called World System: Local Initiative and Local Response. Dialectical Anthropology 2 (4):253–270.
Montenegro, Esther
1988 El Hombre Que Ha Visto Morir el Hato Nacional. Realidad 4 (18):9–10.
Myers, Norman
1981 The Hamburger Connection: How Central America's Forests Become North America's Hamburgers. Ambio 10 (1):3–8.
1990a Primer Lugar en Deforestación en Latinoamérica: Costa Rica Destroza Sus Bosques. 23 September.
Nations, James D.
1992 Xateros, Chicleros, and Pimenteros: Harvesting Renewable Tropical Forest Resources in the Guatemalan Petén. *In* Conservation of Neotropical Forests: Working from Traditional Resource Use, ed. Kent H. Redford and Christine Padoch, 208–219. New York: Columbia University Press.
Nations, James D., and Daniel I. Kromer
1983a Rainforests and the Hamburger Society. Environment 25 (3):12–20.
1983b Central America's Tropical Rainforests: Positive Steps for Survival. Ambio 12 (5):232–238.
Nations, James D., and H. Jeffrey Leonard
1986 Grounds of Conflict in Central America. *In* Bordering on Trouble: Resources and Politics in Latin America, ed. Andrew Maguire and Janet Welsh Brown, 55–98. Bethesda, Md.: Adler and Adler.
Nations, James D., and Ronald B. Nigh
1978 Cattle, Cash, Food, and Forest: The Destruction of the American Tropics and the Lacandón Maya Alternative. Culture and Agriculture 6:1–5.
Navarro, Carlos, and Carlos E. Reiche
1986 Análisis Financiero de una Plantación Familiar de Gliricidia Sepium en Guanacaste, Costa Rica. Turrialba, Costa Rica: Centro Agronómico de Investigación y Enseñanza. Mimeo.
Ortner, Sherry B.
1984 Theory in Anthropology since the Sixties. Comparative Studies in Society and History 26 (1):126–166.
Parsons, James J.
1976 Forest to Pasture: Development or Destruction? Revista de Biología Tropical 24 (supp. 1):121–138.

Partridge, William L.
1984 The Humid Cattle Ranching Complex: Cases from Panama Reviewed. Human Organization 43 (1):76–80.
Patterson, Alan
1990 Debt for Nature Swaps and the Need for Alternatives. Environment 32 (10):4–13, 31–32.
Pearce, Fred
1990 Brazil, Where the Ice Cream Comes From. New Scientist 127 (July 7):45–48.
Peters, Charles M., Alwun H. Gentry, and Robert O. Mendelsohn
1989 Valuation of an Amazonian Rainforest. Nature 339:655–656.
Porras Zúñiga, Anabelle, and Beatriz Villarreal Montoya
1986 Deforestación en Costa Rica. San José: Editorial Costa Rica.
RAN (Rainforest Action Network)
1989 Hamburger Connection Update—June 1989. San Francisco: RAN. Mimeo.
Realidad
1988 Piden Que Salve de la Ruina a los Agricultores. Realidad 4 (18):7–9.
Repetto, Robert
1990 Deforestation in the Tropics. Scientific American 262 (4):36–42.
Roux, Bernard
1975 Expansion du Capitalisme et Développement du sous Développement: L'Integration de l'Amérique Centrale au Marché Mondial de la Viande Bovine. Revue du Tiers Monde 16:355–380.
Rudel, Thomas K.
1989 Population, Development, and Tropical Deforestation: A Cross-National Study. Rural Sociology 54 (3):327–338.
Rutsch, Matilde
1980 La Cuestión Ganadera en México. Mexico City: Centro de Investigación para la Integración Social.
Sader, Steven A., and Armond T. Joyce
1988 Deforestation Rates and Trends in Costa Rica, 1940–1983. Biotropica 20 (1):11–19.
Sáenz Maroto, Alberto
1981 Erosión, Deforestación y Control de Inundaciones en Costa Rica. San José: Universidad de Costa Rica.
Sanderson, Steven E.
1986 The Transformation of Mexican Agriculture. Princeton, N.J.: Princeton University Press.
Sanderson, Steven E., and Ana Doris Capistrano
1991 The Tyranny of the External: Links between International Economic Change and Natural Resource use in Latin America. Paper presented at the Latin American Studies Association, 4–6 April, Washington, D.C.

SEPSA (Secretaría Ejecutiva de Planificación Sectorial Agropecuaria, Costa Rica)
1983 Encuesta Nacional de Ganado Bovino 1982. San José: SEPSA.
1986a Comportamiento de las Principales Actividades Productivas del Sector Agropecuario Durante 1985. San José: SEPSA.
1986b El Sector Agropecuario. San José: SEPSA.

SEPSA et al. (Secretaría Ejecutiva de Planificación Sectorial Agropecuaria, Instituto Interamericano de Ciencias Agrícolas, Ministerio de Agricultura y Ganadería, Banco Nacional de Costa Rica, Federación de Cámaras de Ganaderos, Oficina Nacional de Semillas)
1985 Programa de Reactivación de la Ganadería Bovina en Costa Rica. Mimeo.

Sevilla, Roque
1990 El Canje de la Deuda por Conservación. Los Casos de Bolivia, Ecuador y Costa Rica. *In* Diálogo con Nuestro Futuro Común: Perspectivas Latinoamericanas del Informe Brundtland, ed. Günther Maihold and Víctor L. Urquidi, 139–161. Caracas: Fundación Friedrich Ebert-Editorial Nueva Sociedad.

Shane, Douglas R.
1986 Hoofprints on the Forest: Cattle Ranching and the Destruction of Latin America's Tropical Forest. Philadelphia: Institute for the Study of Human Issues.

Slutsky, Daniel
1979 La Agroindustria de la Carne en Honduras. Estudios Sociales Centroamericanos 22:101–205.

Spielman, Hans O.
1972 La Expansión Ganadera en Costa Rica: Problemas de Desarrollo Agropecuario. Informe Semestral [Instituto Geográfico Nacional] July–December, 33–57.

Stanley, Denise
1991 Demystifying the Tragedy of the Commons: The Resin Tappers of Honduras. Grassroots Development 15 (3):27–35.

Tangley, Laura
1986 Costa Rica—Test Case for the Neotropics. BioScience 36 (5):296–300.

Thrupp, Lori Ann
1989 The Political Ecology of Natural Resource Strategies in Central America: A Focus on Costa Rica. Paper prepared for the Latin American Studies Association Conference, 3–6 December, Miami, Fla.

Tosi, Joseph A.
n.d. Los Recursos Forestales de Costa Rica. San José: Centro Científico Tropical. Mimeo.

Ugalde Arias, Luis A., and Hans M. Gregersen
1987 Incentives for Tree Growing in Relation to Deforestation and the

Fuelwood Crisis in Central America. Turrialba, Costa Rica: Centro Agronómico Tropical de Investigación y Enseñanza. Mimeo.
Uhl, Christopher, and Geoffrey Parker
1986a Our Steak in the Jungle. BioScience 36 (10):642.
1986b Is a Quarter-Pound Hamburger Worth a Half-Ton of Rain Forest? Interciencia 11 (5):210.
Utting, Peter
1991 The Social Origins and Impact of Deforestation in Central America. Discussion paper 24. Geneva: United Nations Research Institute for Social Development.
Vargas Mena, Emilio
1990 Ecología y Deuda Externa. Aportes 66:12–14.
Wallerstein, Immanuel
1974 The Modern World-System: Capitalist Agriculture and the Origins of the European World Economy in the Sixteenth Century. New York: Academic Press.
Williams, Robert G.
1986 Export Agriculture and the Crisis in Central America. Chapel Hill: University of North Carolina Press.
WRI (World Resources Institute)
1985 Tropical Forests: A Call for Action. Washington, D.C.: WRI.
WRI, UNEP, and UNDP (World Resources Institute, United Nations Environment Programme, United Nations Development Programme)
1990 World Resources 1990–91. New York: Oxford University Press.

Chapter 2

Development, Rural Impoverishment, and Environmental Destruction in Honduras

Susan C. Stonich

The expansion and diversification of export agriculture have been the bases of development efforts in Central America since the end of World War II. The growing global concern over the environmental consequences emanating from the implementation of such development strategies throughout the Third World has raised important questions about the complex interrelationships between social processes and the natural environment (Bruntland 1987). These concerns point to the need to link specific development efforts systematically with their environmental consequences. This chapter confronts such a requirement by centering on those aspects of contemporary natural-resource-based development in Central America that have significant ramifications on the environment. The study integrates environmental factors into political-economic analysis to demonstrate the systemic linkages among social processes emanating from this prevailing development model and environmental decline.[1]

The chapter begins with a discussion of how environmental questions were incorporated into political-economic analysis. This is followed by a summary of development efforts to promote nontraditional agricultural exports in Central America since World War II. Then, in order to demonstrate the continuing complex, multilevel articulation among social, economic, and environmental factors, the chapter focuses on the impact of the most recent trend in expanding exports in one region of Central America—southern Honduras—where the growth of nontraditional exports has been significant and that is currently distinguished by heightened social differentiation, deepened human impoverishment, and escalating environmental destruction (figure 1). This chapter concentrates on the most important of several nontraditional exports recently promoted in lowland and coastal areas along the Gulf of Fonseca—especially cultivated shrimp and irrigated melons. The integrated approach shows the interconnections among the dynamics of agricultural development, the associated patterns of capital accumulation, the worsening inequality and impoverishment, and the serious problems of environmental degradation. This chapter illustrates how larger international and national forces have affected both the people and the natural environment and how, in turn, the local people are struggling to shape and mitigate those forces and to affect change beyond the local level.

Political Ecology

The interrelationships between social processes and environmental destruction are revealed by examining the multilevel mechanisms involved in expanding and diversifying agricultural exports, their relationship to changes in access to and use of resources, and their impact on the local and regional economy and ecology. This approach expands the perspective of political economy to include the distribution and use of natural resources and the dynamic contradictions between society and natural resources. This integrated perspective, termed *political ecology,* has been used in a variety of disciplines to demonstrate how interconnected social, economic, and political processes affect the way natural resources are exploited.[2]

The chapter demonstrates how international development organizations, multinational corporations, the state, and individuals, by focusing on short-term needs, created extremes of wealth and poverty that exacerbated resource abuse. The uneven growth of export production is also shown: as the production of agricultural exports expands, an escalating percentage of total product comes from larger producers with superior access to credit, technology, and markets, while a growing number of smaller producers become displaced and destitute. The ecological and environmental consequences of the particular commodities being promoted have expanded and accelerated the destruction of highland, lowland, and coastal zones.

The conclusions apply to much of Central America, where there is overwhelming evidence that, because of economic and population pressures, governments and individuals are overexploiting the natural re-

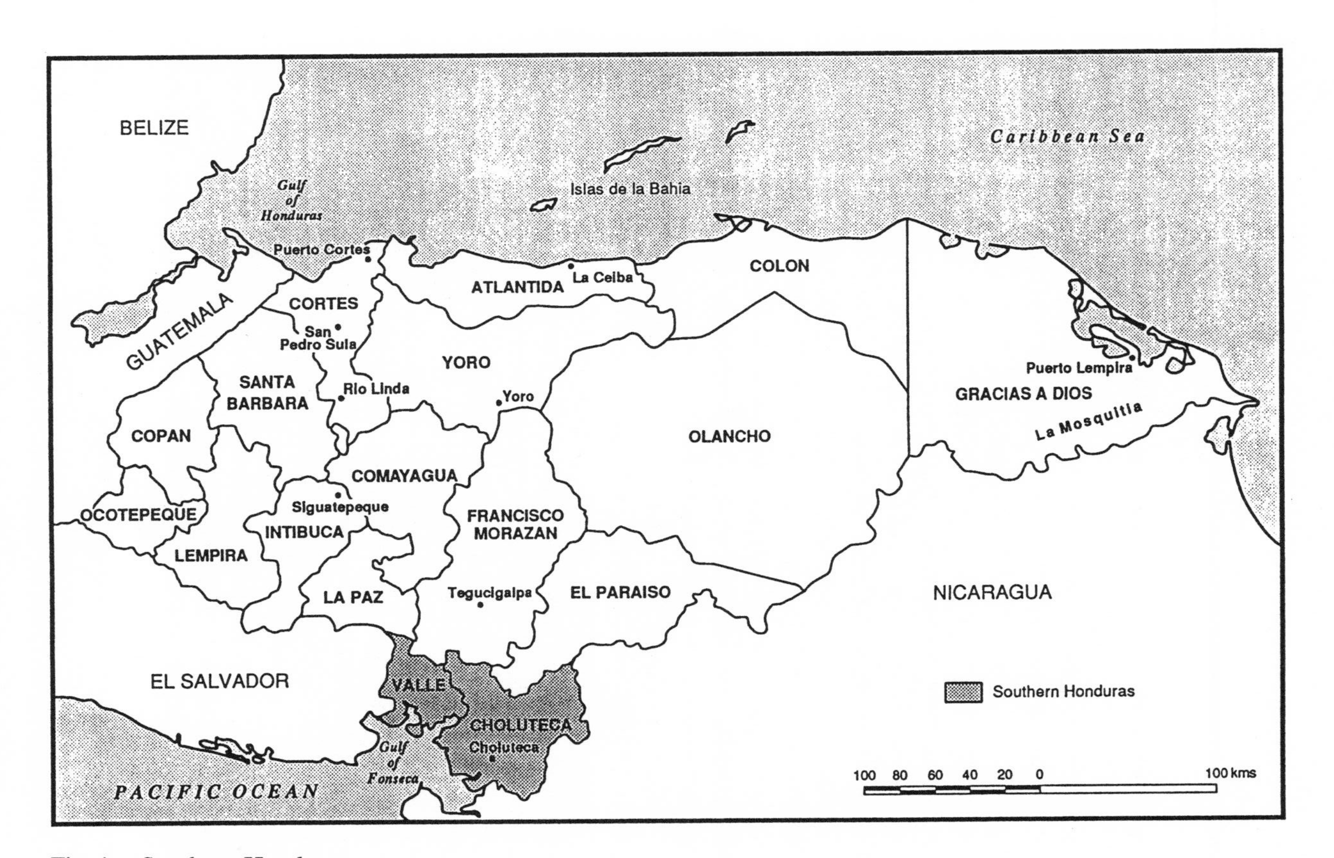

Fig. 1. Southern Honduras

sources that they control in order to generate income to satisfy immediate needs—whether those needs are to generate foreign exchange at the national level or to increase current income at the household level.

Agricultural Diversification and Export-led Growth

Development schemes aimed at alleviating Central America's social and economic problems have, historically, stressed intensified exploitation of the region's natural resources through augmented exports of agricultural commodities and forest products, enhanced agricultural productivity, and expanded industrial fisheries. There is plentiful evidence documenting how succeeding waves of export expansion have displaced small farmers from their land, often initiating cycles of violence and repression (Durham 1979; Williams 1986; Brockett 1988). Social conflict has been somewhat ameliorated in the past, however, by the hope of migration to the agricultural frontier, by the limited presence of domestic markets for the products of peasant agriculture, and, especially during the 1980s, by massive amounts of foreign assistance.[3]

Since World War II, mounting political unrest in Central America, perceived as a threat to economic and national security, has stimulated U.S. policy initiatives that emphasize economic development along with military assistance. The Alliance for Progress, begun in 1961, was a multibillion-dollar development effort built on export diversification and the expansion of agricultural export earnings, especially cotton, sugar, and beef—the "nontraditional" agricultural exports of that era. Export quotas to the United States were expanded, efforts were undertaken to stimulate trade and to modernize production, and credit programs were instituted to augment the production of nontraditional commodities. These initiatives were directed into the region through the expanded investment of multinational corporations and through large landowners, merchants, and industrialists, who formed the elites of each country (Stonich and DeWalt 1989). Evaluated in terms of macroeconomic indicators, the Alliance for Progress appeared successful, but economic gains were not enjoyed by the majority of the population: land ownership became more concentrated and control of production became dominated by large-scale national and international investment (Williams 1986). The inequitable distribution of wealth in the region was exacerbated as a relatively small elite benefited disproportionately from the economic growth of the period (Bulmer-Thomas 1987).

During the 1970s, because of a combination of internal and external factors (as declines in international commodity prices, soaring costs for oil and such technological inputs as pesticides, the collapse of regional markets, and accelerated costs of further investment and debt financing) the region sank into a severe economic crisis, accompanied by worsening inequality (Perez-Brignoli 1982; ECLAC 1986; Bulmer-Thomas 1987). By 1980, 63.7 percent of Central Americans (76.2 percent in rural areas) were living in poverty, without sufficient income to cover essential needs, such as food, housing, clothing, and basic services (ECLAC 1986: 20). The effect was especially severe on the poor, and dissension and repression grew in tandem with the intensified economic crisis (Williams 1986; Brockett 1988).

In response to deteriorating social, economic, and political conditions in the early 1980s, U.S. policymakers proposed economic reforms, in order to defuse the various crises in the region. The strategy was implemented on 1 January 1984, when the Caribbean Basin Initiative (CBI) went into effect, and was made permanent in August 1990 with the passage of the Caribbean Basin Economy Recovery Expansion Act.[4] The goal was to establish regional stability through economic growth, stimulated by domestic and foreign investment. Agricultural diversification through the promotion of a new set of nontraditional export crops was central to this scheme (Paus 1988). Nontraditional export crops included many fresh, frozen, processed, and otherwise preserved fruits and vegetables (e.g., melons, miniature papayas, mangos, snow peas, broccoli, eggplant); root crops; edible nuts; live plants and cut flowers; and the most commercially desirable species of crustaceans and mollusc—especially shrimp and lobster (Paus 1988). In a context of stagnating traditional exports, mounting debt, and worsening poverty, and linked to structural adjustment programs imposed by the World Bank, the International Monetary Fund, and others, the objectives of nontraditional growth were to broaden export earnings while facilitating the servicing of the region's foreign debt (Murray and Hoppin 1992).[5]

Central American governments are currently augmenting the nontraditional export sector and seeking ways to diversify their export base to generate larger amounts of foreign exchange. Foreign exchange earnings from nontraditional agricultural and nonagricultural exports have risen from U.S. $423 million in 1983 (approximately 12 percent of total export earnings from the region) to U.S. $1.3 billion in 1990, and have

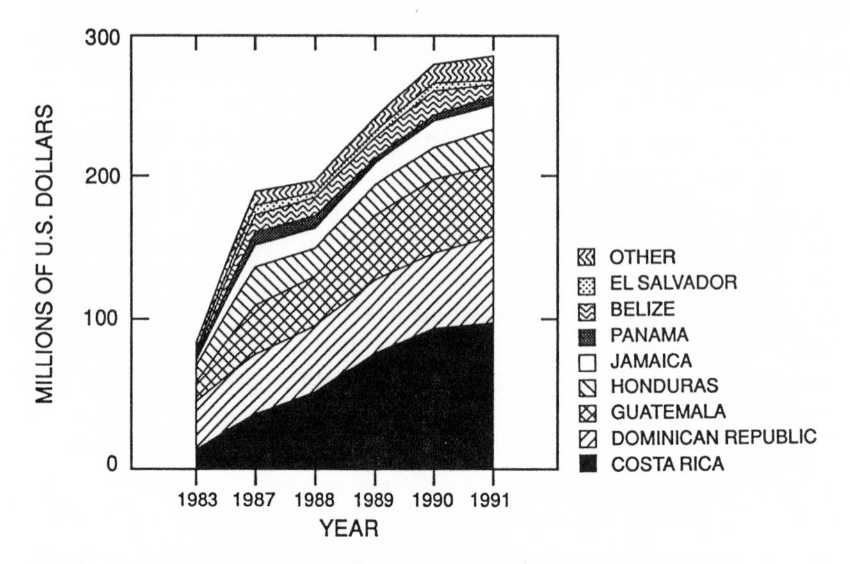

Fig. 2. U.S. imports of nontraditional products from Caribbean

the potential of reaching U.S. $4 billion by 1996—comprising about 50 percent of total export earnings (USAID 1991b).

As shown in figure 2, U.S. imports of nontraditional horticultural products (not including fish and shellfish) from CBI countries increased from U.S. $88 million to U.S. $288 million (227 percent) between 1983 and 1991 (USDA 1992a). During that time, Costa Rica was the largest CBI beneficiary, accounting for 27.6 percent of nontraditional imports to the United States, followed by the Dominican Republic with 23.7 percent, Guatemala with 17.4 percent, Honduras with 10.9 percent, and Jamaica with 6.3 percent (computed from data in USDA 1992a). The other Central American countries (Belize, Panama, El Salvador, and Nicaragua) supplied a total of approximately 10 percent of nontraditional imports during the same period.[6]

In 1991, fresh melons (cantaloupe, watermelons, and winter melons) were the most important nontraditional horticultural import commodities, followed in rank by fresh pineapple, prepared vegetables, fresh peas and beans, pineapple juice, and yams, which together accounted for 56 percent of horticultural imports, excluding bananas and plantains, from the region (USDA 1992a). The significant recent growth in melons imported through the CBI to the United States is shown in figure 3. Between 1987 and 1991, the quantity of melons imported into

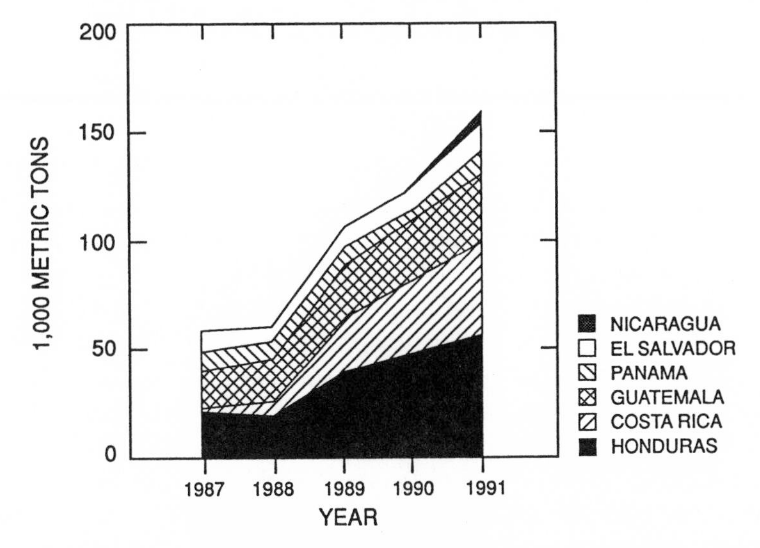

Fig. 3. Quantity of melons imported into the U.S. from Central America, 1987–91. (Data from *Horticultural Products Review,* Foreign Agricultural Service, Circular Series, Supplement 1-92, March, USDA, 1992.)

the United States grew 173 percent—from 59,000 to 156,000 metric tons—with the major suppliers being Honduras, Costa Rica, and Guatemala (computed from USDA 1992b).

In addition to a variety of horticultural products, Central American countries are also encouraging the exploitation of marine resources, especially shrimp mariculture, as a principal means of attacking the region's continuing economic crisis. Several international agencies, including the UN and USAID, predict that shellfish (primarily cultivated shrimp) will be the most important primary nontraditional export commodity from the region during the 1990s and are promoting the expansion of shrimp mariculture in coastal zones (Stonich 1991b, 1992a). During the 1980s, shrimp exports from Central America relied less on capture fisheries (largely because overfishing and destruction of habitats significantly reduced catches) and more on shrimp mariculture, primarily located along the Pacific Coast. Honduras, Panama, and Costa Rica led the region in the expansion of cultivated shrimp production (computed from country-level unpublished data). The importance of shrimp as an export crop is demonstrated by its unprecedented growth in Honduras, where, by 1987, foreign exchange earnings from shrimp (most produced on farms) were surpassed only by export earnings from bananas and coffee (Stonich 1991b, 1992a).

The Honduran Economy

Duplicating patterns that occurred throughout Central America, the Honduran real Gross National Product, which grew by 4.5 percent annually between 1967 and 1977, increased by only 1.9 percent annually from 1977 to 1987 (World Bank 1990)—a rate that was well below the annual population growth rate of 3.4 percent for the same period. Albeit agriculture performed better than other sectors of the economy (the agricultural growth rate of 2.2 percent was the highest of all Central American countries), per capita growth in fact decreased and a general decline in wages and in the standard of living followed (ADAI 1987). The purchasing power of rural households declined more than that of urban households in part because of escalating rural unemployment and declines in agricultural prices in relation to nonagricultural prices. As a result, average nutritional levels were lower in the late 1980s than in 1970; the average energy deficit in rural areas was approximately 20 percent, and 38 percent of Honduran children under the age of five exhibited some degree of malnutrition (Stonich 1991c). In spite of escalating food deficits and a growing dependence on international food aid, the Honduran government continues to promote the expansion of export agriculture in order to generate foreign exchange (Stonich 1991c).

Economic liberalization was an essential element of the platform of the National Party that came into power in 1989. One of President Rafael Leonardo Callejas's first actions was to devalue the Honduran currency—long a demand of many exporters; another was to declare the nation bankrupt. Given the philosophy of the ruling party and the almost total economic collapse, it is unlikely that the national government will redirect its policies away from attempting to expand export production.

Notwithstanding the above constraints, conserving Honduras's natural resource base is crucial because the country remains exceedingly dependent upon renewable natural resources to generate income in the predominant economic sector—agriculture, forestry, and fisheries. In 1988, 61 percent of the population persisted in living in rural areas and were employed in the agricultural sector (World Bank 1990). Throughout the 1980s natural-resource-based commodities continued to be the principal means of earning foreign exchange—generating more than 80 percent of export earnings (World Bank 1982–90).

Southern Honduras: The Region

Southern Honduras is a triangular-shaped region (6,840 km^2) located in tropical dry and subtropical moist forest zones in the Pacific watershed of Central America (Holdridge 1962) (fig. 1). The area that extends along the Gulf of Fonseca is covered by a band of mangrove and marsh grass, beyond which lies one of the infrequent broad plains on the Pacific coast of Central America. Savanna gives way to steep foothills, which rapidly become the jagged mountain ranges to the northeast that comprise 62 percent of the region. These volcanic mountains seldom reach elevations of more than 1,600 meters, but they are exceedingly rugged and fashion innumerable segregated valleys. The region is marked by distinct dry and rainy seasons with unpredictable precipitation patterns, and soils that are extremely prone to erosion. As a result, agriculture is very risky and the area is highly vulnerable to environmental degradation (USAID 1982; CRIES 1984; Stonich 1986; SECPLAN/USAID 1989). Intensifying this susceptibility to environmental degradation is the uneven spatial distribution of the growing human population (Stonich 1989).

Growth and diversification of agricultural production distinguished the southern Honduran economy from the 1950s to the 1980s as a sequence of export commodities were fostered, including cotton, sugar, livestock, sesame, and, most recently, nontraditionals, such as cultivated shrimp and melons. For the government, in persistent fiscal crisis, contending with the repayment of expanding external debt has been more important than sustaining natural resources. Cotton, cattle, sugar, melons, and shrimp bring on international assistance and help satisfy the necessity for foreign exchange, regardless of their effect on the natural environment.

The Cotton Boom

Although the transformation of forest to pasture that began in Central America in the 1960s as part of the livestock boom is well known, the ecological outcomes of the cotton boom that began in the 1950s along Central America's Pacific coastal plain were no less significant. Moreover, it was cotton cultivation that began the transformation of traditional social patterns of production in the south (Stares 1972:35; Durham 1979:119; Boyer 1982:91). In the process of creating large cotton estates, large landowners drove many small farmers from their land, both legally and illegally. They revoked peasant tenancy or sharecropping rights, raising rental rates exorbitantly, and evicting peasants forcibly from national land or from land of unclear tenure (K. Parsons 1975; White 1977; Durham 1979; Boyer 1982). A major effect (magnified later by the expansion of cattle production) was to exacerbate inequalities in access to land and displace many poor farmers from more suitable agricultural land in the south.

Although cotton had been grown in the area since preconquest times, large-scale commercial cultivation of cotton was introduced in the late 1940s and early 1950s by Salvadorans who brought seeds, chemicals, machinery, and their own labor force to the area. Aggressive Salvadoran farmers secured Honduran bank loans, rented (or purchased) large tracts of land from Honduran owners, and began commercial production. They were joined by Honduran farmers who first began producing on a minor scale but who by 1960 expanded production and formed their own ginning and marketing cooperative. When the Salvadorans were expelled from the country after the Salvadoran-Honduran "Soccer War" in 1969, their property was confiscated and became available to the Honduran growers (Durham 1979; Stonich 1986:118).

As in El Salvador and Nicaragua, commercial cultivation involved the combination of mechanization (in land preparation, planting, cultivation, and aerial spraying) and manual labor (in harvesting). Cotton cultivation along the Pacific coastal plain is highly mechanized and dependent on the heavy use of chemical inputs. The indiscriminate use of pesticides in the cotton growing regions remains one of the most pervasive environmental contamination and human health problems throughout Central America. Water from cotton growing areas of southern Honduras shows heavy use of DDT, Dieldrin, Toxaphene, and Parathion (USAID 1982) and the results of a 1981 study to determine the levels of pesticide poisoning in the Choluteca area revealed that approximately 10 percent of the inhabitants had pesticide levels sufficiently high to be considered cases of intoxication (Leonard 1987:149). A more recent study of pesticide contamination of surface and ground water in Choluteca found high levels of organocholorine pesticides, including DDT, used in cotton production for more than thirty years. Although most of these pesticides are no longer commonly used in the region, their effects linger (Buseo et al. 1987). In sum, the ecological costs of expanded cotton production along the Pacific coastal plain of Central America—including southern Honduras—have been exorbitant: fertile soils have been depleted through continuous cultivation; soil erosion has been accelerated because of the general lack of planting leguminous ground cover capable of absorbing rainfall from the heavy thunderstorms that occur during the rainy season; the growth of harmful insects has been encouraged because of previous strategies aimed at the total eradication of cotton pests with unregulated applications of DDT and other chlorinated insecticides; and land and water contamination, as well as high levels of pesticide residues in food supplies, threaten human health (ICAITI 1977; Weir and Shapiro 1981; Bull 1982; Botrell 1983; Boardman 1986; Williams 1986; Leonard 1987).

The amount of land planted in cotton, the amount produced, and the number of jobs in the industry (estimated to be 76.7 person-days per hectare per year) fluctuated considerably between the 1950s and the 1980s. Mounting cotton prices during the 1960s sustained the escalating costs of production brought about by spiraling pesticide and fertilizer use. By 1965, cotton production reached a peak of 33,000 metric tons, the area under cultivation climbed to 15,000 hectares, and the number of jobs grew to 13,000. Declining world prices, severe infestations of pesticide resistant insects, and the disorder brought about by the "Soccer War" abruptly ended the cotton boom, and thousands of farmers were forced into bankruptcy or into the cultivation of other crops, while thousands of harvesters were obliged to look for other employment. By 1970, cotton production fell to 9,000 metric tons and the land area planted in cotton declined to 4,000 hectares. Improved market conditions in the late 1970s stimulated cotton cultivation once again, and the area in cultivation climbed to 13,000 hectares (total production of 23,000 metric tons) by 1980 (ECLAC 1987; FAO-PY 1989). At that time, declining international market conditions, social unrest in areas bordering El Salvador and Nicaragua, and increasing losses to chemically resistant pests combined to undermine the Honduran cotton industry once again. By 1988, only about 1,200 hectares of land remained planted in cotton—approximately the same as at the start of the cotton boom in the 1950s. The social impact of the latest collapse was significant and was exacerbated by the simultaneous decline in the livestock and sugar markets, other important commodities grown in the region.

The Cattle Boom

Opportunities for the production of other export commodities expanded concurrently with the decline of cotton. Between 1960 and 1983, 57 percent of the total loan funds allocated by the World Bank for agriculture and rural development in Central America promoted the production of beef for export (Table 4–1 in Jarvis 1986:124). During that same period, Honduras obtained 51 percent of the total World Bank funds disbursed in Central America—of which 34 percent were for livestock projects (Table 4–1 in Jarvis 1986:124). In a context of declining agricultural commodity prices, high labor costs, unreliable rainfall, and international and national support for livestock, landowners reallocated their land from cotton and/or grain cultivation to pasture for cattle. Land reform programs also encouraged investment in livestock: landowners who feared expropriation of unutilized land fenced it, planted pasture, and stocked it with cattle as a way of establishing use without increasing labor inputs (Jarvis 1986:157). Between 1947 and 1988, the number of head of cattle rose 150 percent throughout Central America and 220 percent in Honduras (computed from FAO-PY, various years).

The growth in livestock production in southern Honduras between the 1960s and the early 1980s was unprecedented (Stonich 1986:139–43). Expansion took place not only in the lowlands and foothills where cattle raising traditionally occurred, but also in the highlands, where many of the larger peasant farmers augmented cattle production. Because of the land-extensive system of cattle raising in the south (1 hectare per head) and the low labor demands (6.3 person-days per hectare per year), the spread of cattle raising could not absorb the growing landless and land-poor rural population and, in fact, intensified their expulsion (Howard-Ballard 1987).

Throughout Central America, the environmental corollary of the cattle boom was the reallocation of land from forest, fallow, or food crops to pasture (see J. Parsons 1976; Myers 1981; DeWalt 1983; Nations and Komer 1983; Shane 1986; Williams 1986). In southern Honduras, between 1950 and the mid-1970s, pine and deciduous forests decreased by 44 percent and fallow land (vital to regenerative shifting cultivation systems) plunged by 58 percent, while pasture land rose by 53 percent (Stonich 1989). By 1974, deciduous and pine forests covered 13 percent of the region while pasture comprised over 60 percent of the total land area.

Simultaneously, the number of farms that grew the basic food crops (corn, sorghum, and beans), the total number of hectares of land used to produce those crops, and total and per capita production stagnated or declined (Stonich 1989, 1991c). These trends continued throughout the 1980s: between 1950 and 1990, the area in cultivation of basic grains plummeted: corn by 51 percent, sorghum by 10 percent, and beans by 86 percent. Per capita production also tumbled: corn by 72 percent, sorghum by 64 percent, and beans by 95 percent.[7]

The Social Consequences of Development

The social consequences of the expansion of cotton and cattle production on rural areas of the south have been discussed in great detail elsewhere (see R. White 1977; Durham 1979; Boyer 1982; Stonich 1986, 1989, 1991a; Stonich and DeWalt 1989). Land ownership became more concentrated, which augmented social differentiation and impoverishment. Between 1952 and 1974, farms of less than five hectares (the estimated minimum amount of land needed to generate household income sufficient to meet basic needs) increased from 60 percent to 68 percent of all farms. This growth was particularly significant in farms of less than one hectare, which increased from 12 percent to 21 percent of total farms. By 1974, an estimated 34 percent of southern Honduran families were landless (Stonich 1986:143). More recent community level studies reveal that more than 80 percent of rural highland families are landless or land-poor (Stonich 1993).

The concentration of landholdings increased the dependence of the majority of families on off-farm income from petty commodity production, wage labor, migration, and remittances from family members. Analyses of household budget studies indicated that off-farm incomes contributed from 20 percent to 60 percent of total incomes and, not surprisingly, farm size was inversely related to the degree of household dependence on off-farm income (Stonich 1991a). Thus, while the emergence of a class of "rich" peasants was founded on prospects for accumulating capital, land-poor and landless peasants, unable to produce adequately to satisfy their own needs, also became integrated into the capitalist agricultural sector, but under conditions of physical marginalization and subordination to landed, commercial, and agro-industrial firms. Farmers with larger holdings, who had little incentive to produce basic food grains for the market, persisted in acquiring land, mechanizing production where possible, and expanding livestock production (Stonich and DeWalt 1989).

Among the corollaries of the grim economic situation was widespread undernutrition. The national planning agency estimated that by the mid- to late-1970s, 41 percent of all southern Honduran families did not meet minimum nutritional requirements (Stonich 1986:152–154). Data collected in 1982–83 in nine highland and lowland communities showed that 65 percent of children under sixty months of age were stunted. Further, there was a close relationship between access to land and nutritional status—of all undernourished children under sixty months of age, 70 percent belonged to landless families and 15 percent to families with access to less than two hectares (Stonich 1991c).

In the absence of effective programs to redistribute land (Ruhl 1985), impoverished families from the south were spurred to leave the region, beginning in the 1960s and accelerating their pace during the 1970s and the 1980s (Stonich 1989). Stimulated to seek land and/or work, migrants from the south have contributed to the growth of the urban centers of Tegucigalpa and San Pedro Sula and have been the foremost colonizers of the humid tropical areas of Olancho and the Mosquitia (Stonich 1991a). Between 1974 and 1987, the urban growth rate in Honduras was 5.6 percent, much higher than the overall population growth rate of about 3.4 percent during the same period (USAID 1989b). The squalid slums bordering urban centers attest to the environmental problems caused by this rural-to-urban migration. In addition, migrants from degraded areas in the south comprise a significant percentage of new settlers to the tropical humid forests of the north and northeast. Deforestation has taken a heavy toll, as newly arriving colonizers clear forest for crops, cattle, and fuelwood while simultaneously encroaching on the land inhabited by Honduras's small remaining indigenous population.

Local Ecological Responses to Regional Processes

The changes in agricultural systems that accompanied the agrarian transformation included the emergence of potentially destructive agricultural practices on the part of farmers having landholdings of all sizes. Table 1, based on national-level statistics, associates a number of such practices to land tenure. It shows that the percentage of land in cultivation is inversely related to the size of landholdings, while the percentage of land in pasture, the percentage of total cattle owned, and the mean number of cattle owned are positively related to the size of landholdings. It also offers another way to measure agricultural intensification—through the use of purchased and/or industrial inputs. It reveals that the larger the farm, the higher the percentage of farmers utilizing such inputs. These national-level data disclose two broad patterns in agricultural adaptation. Larger, more commercially oriented farms became more capital intensive, using expanded access to agricultural credit and advanced agricultural technologies in order to secure higher yields from the same land area. At the same time, these farms expanded the land-extensive system of cattle ranching. On the other hand, as a result of escalating population densities and declining incomes, rural households with diminished holdings were forced to utilize land more intensively, by cultivating high proportions of their land and by shortening the fallow period. However, as indicated by table 2 (computed from community-level studies), the dichotomy in agricultural practices suggested by the aggregate census data is an oversimplification. Although it, too, shows comparable relationships among the size of landholdings, the percentages of land in cultivation and in pasture, and the ownership of cattle; it also reveals that small farmers made considerable use of purchased seed and industrial inputs.

The environmental consequences of these various actions are extremely important. The agricultural practices of each socioeconomically differentiated subgroup (i.e., rich farmers, medium farmers, small farm-

TABLE 1. Comparison of Honduran Farms by Size Category

Farm Size	% Total Farms	% Total Farmland	% Land in Cultivation[a]	% Land in Pasture[b]	% Total Cattle	Mean No. Cattle	% Using Inputs[c]
<1	17	1	96	2	1	4.0	
1–5	47	8	70	12	9	5.7	9.5
5–10	14	9	42	27	8	8.7	16.0
10–20	10	10	30	37	10	13.5	22.1
20–50	8	17	20	49	17	25.1	31.7
50–100	2	11	14	60	12	55.6	54.1
>100	2	44	9	66	42	237.1	
Total	100	100	22	51	100	21.7	15.2

Source: Data from DGECH 1976 and USAID 1982.

[a]Includes cultivation of annuals and perennials.

[b]Includes natural pasture, improved natural pasture, and cultivated pasture.

[c]Includes fertilizer, insecticides, fungicides, herbicides, veterinary products, and purchased seeds.

ers, and renters) resulted in mutually reinforcing destructive effects on the biophysical system (Stonich 1989). Although pasture is predominant on most of the less steep land that is owned by large farmers, poor land management practices resulted in much of this land being overgrazed. On the other hand, small farmers, as a result of land fragmentation, declining incomes, and the subsequent reallocation of household labor, increased purchases of industrial inputs, expanded the percentage of their on-farm production sold for cash, and adopted new marketing arrangements. Poor farmers continued to clear steeper and more mar-

TABLE 2. Agricultural Practices by Land Tenure Arrangements, Highland Village Data, Southern Honduras, 1983.

Tenancy		Land Use and Cattle						
Tenancy (ha.)	No.	% Land in Cultivation[a]	% Land in Pasture	Length Fallow (years)	Mean No. Cattle (range)	% Using Purchased Seed[b]	% Using Insect[c]	% Using Herb.[d]
Renters[e]	74	95	—	2.7	.17 (0–4)	64	7	—
Owners[f]								
<1	23	80	—	2.7	.22 (0–3)	61	13	30
1–5	87	51	4	3.2	.22 (0–3)	45	32	28
5–20	15	23	21	3.8	2.5 (0–13)	20	20	7
20–50	5	6	48	5.0	8.0 (7–9)	20	—	—
>50	1[g]	6	20	6.0	50 (50)	—	100	—

Source: Author's calculations based on unpublished survey conducted by author. Details on survey data, methodology, and results available from author.

[a]Includes the major food crops—maize, sorghum, and beans.

[b]Fifty-two percent of farmers had purchased seed (maize, sorghum, or beans) in order to resow their fields because drought conditions at the beginning of the agricultural cycle resulted in the loss of their initial planting. Seventy-seven percent of resown maize and 35 percent of resown sorghum was purchased.

[c]Approximately two-thirds of farmers treated seeds before planting—a process that involved nothing more than putting the seeds into the insecticide mixed with a little water and stirring them around with bare hands. Insecticides were also applied to crops growing in the field. Malathion and Diptheryx were the two most commonly used. They were usually applied with a backpack applicator.

[d]Most farmers weeded twice. If herbicides were used they were applied for the first weeding. The second weeding usually was done with machetes. Few farmers were aware of the herbicide they were using, but the two most common varieties were 2-4-D and Herbisol.

[e]Mean area of rented land, 1.4 hectares.

[f]Fifty-one percent of such owners also rented land.

[g]Largest landowner rents additional grazing land in lowlands.

ginal land, were less likely to maintain soil conservation measures, persisted in destructive burning, and shortened fallow periods, while being drawn directly into the forest-to-pasture conversion process by larger landholders (Stonich 1989, 1993).

Throughout the highlands these practices have contributed to extensive deforestation and land degradation—loss of soil fertility, erosion, and landslides (USAID 1982; CRIES 1984; Stonich 1986, 1989; SECPLAN/USAID 1989). Widespread watershed deterioration has accelerated the siltation of rivers and of mangrove areas along the coast. Increased sediment loads have mixed with pesticide-contaminated runoff from large farms located in the Pacific lowlands. Together with the substantial harvesting of mangroves for fuelwood in the Pacific coastal salt-extraction industry, for tanbark, and for the expansion of shrimp farms, these diverse factors have augmented the destruction of mangrove forests and threatened coastal fishing grounds.

The New Nontraditionals: The Shrimp and Melon Boom

With the weakening of other economic sectors in the 1970s, the Honduran government began initiating policies to enhance nontraditional export growth in the early 1980s. It declared 1987 as "the year of the exports" and undertook several measures to stimulate private investment and export production, especially of nontraditionals: import taxes for inputs used in export products were eliminated, exporters were allowed to keep part of their export earnings for direct purchases, and investment policy and export regulations were simplified (USAID 1989a). The World Bank, USAID, and the European Commission (EC), among others, encouraged the shift to nontraditionals through the infusion of new projects, loans, and funding (Heffernan 1988). With the financial support of USAID, a number of quasi-official organizations designed to promote Honduran exports through the creation of information and business networks were established: the Honduran Federation of Agricultural and Agro-Industrial Producers and Exporters (FEPROEXAAH), the Foundation for Entrepreneurial Research and Development (FIDE), and the Honduran Agricultural Research Foundation (FHIA).

Between 1980 and 1987, the value of nontraditional agricultural exports (including agricultural crops, agro-industrial products, and shrimp) grew from U.S. $65.7 million to U.S. $107.8 million (USAID 1989c). During that time period, the value of nontraditional agricultural crops rose from approximately U.S. $21 million to U.S. $32 million and the value of exports of shrimp and lobster more than doubled (from U.S. $23.4 million to U.S. $50 million). By 1987, income from the export of shrimp ranked third, after bananas and coffee, in total export earnings for Honduras, supplanting the position that beef exports had previously held (Stonich 1991b), and by 1988, the value of shrimp exports was estimated to be approximately U.S. $78 million (UN 1991). Although nontraditional agricultural exports are being promoted in other areas, lowland and coastal zones along the Gulf of Fonseca in southern Honduras are the sites of two of the most important—irrigated melons and cultivated shrimp (Stonich 1991b). Continuing the pattern established in the postwar period, the Honduran government, in collaboration with the United States and other foreign interests, is attempting to stimulate shrimp and melon exports through improvements in the regional infrastructure, including the construction of an airport for the city of Choluteca, the repair and improvement of the Pan American Highway, and the construction of two industrial parks in the urban centers of Choluteca and San Lorenzo (Stonich 1993).

The Shrimp Boom

The declining importance of beef as an export commodity by 1980 occurred concurrently with the commencement of cultivated shrimp as the most important new export product. Principal investors included transnational corporations and government and military leaders, as well as consortiums of private investors. As in the rest of Central America, the enlargement of shrimp farms was financed by national and international, private and public capital (USAID 1985; SECPLAN/USAID 1989). This included direct financing through loans (e.g., between 1986 and 1989 USAID provided U.S. $7 million for seven large farms, and the quasi-governmental Rural Technologies Program channeled an additional U.S. $1 million to medium-scale producers) and indirect funding in the form of incentives to foreign investors (e.g., through certificates of export promotion and foreign exchange and bonds for tax payments) (USAID/FEPROEXAAH 1989). The total production of cultivated shrimp grew from 130 metric tons to 2,225 metric tons (1611 percent) between 1978 and 1988, and the area in production grew from 1,450 hectares to 5,500 hectares (280 percent) in the three-year period from 1986 to 1989 (USAID/FEPROEXAAH 1989). By 1988, production and exports from shrimp farms exceeded that from industrial fisheries (SECPLAN/USAID 1989). According to estimates by USAID, the area in shrimp farms may expand to more than 15,500 hectares by 1995 with an estimated export value of U.S. $100 million (USAID/FEPROEXAAH 1989).

The environmental analysis of the USAID-funded Investment and Export Development Project asserts that the "pitifully little research on the natural resources of the Gulf of Fonseca's estuaries, mangrove forests, and mudflats" makes it impossible to evaluate adequately the significance of environmental changes emanating from the ongoing expansion of shrimp farms (Castañeda and Matamoros 1990). Without such information it is impossible to predict precisely the consequences of further loss of mangroves, increased sediment load in the water column, construction of roads through the estuaries on levees, instead of bridges, and altered oxidation reduction potential from the loss of the 14,000 hectares of mudflats that have already been consigned and are scheduled for conversion to shrimp farms by 1995 (Castañeda and Matamoros 1990).[8]

The Distribution of Costs and Benefits at the Local Level

The unequal distribution of coastal resources associated with concessions granted by the government agency with that charge, the Honduran Department of Tourism (SECTUR), is evident: of the fifty-seven concessions granted by 1988, twenty-five (44 percent) were small holdings (1–70 hectares) with access to only 723 hectares of land (2.6 percent of total concessions), while the eight largest concessionaires had rights to 19,535 hectares (69 percent) (SECPLAN/USAID 1989). According to a study done by economists at the Honduran National Autonomous University (UNAH), by 1991, five farms owned or had concessions of approximately 1,000 hectares or more: Granjas Marinas (5,055 hectares); Aquamarina Chismuyu (3,000 hectares); Aquacultivos de Honduras (1,540 hectares); Aquacultura Fonseca (957 hectares); and Cumar (934 hectares). Of these, only Aquamarina Chismuyu had a concession on what had been private land; the rest were national land. This same study estimated that approximately 72 percent of the total coastal land utilized for shrimp farms was national land (Banegas Archaga et al. 1991).

The unequal distribution of shrimp farms in operation is clear as well. By 1990 approximately seventy-six farms operated in the south. Of these, five were large farms of more than 250 hectares each (total of 4,200 hectares), twenty-one were medium farms ranging from 20 to 250 hectares in size (total of 1,550 hectares), and fifty were small family farms of less than

20 hectares (total of 250 hectares). The five largest farms (6 percent of the total farms) controlled 70 percent of the land in production, the twenty-one medium farms (28 percent of all farms) managed 26 percent of the farmland, and the fifty smallest farms (66 percent of all farms) had access to only 4 percent of the land in production (Mejia 1991). The rationing of concessions and permits to build ponds, as well as the large capital outlays required for intensive (and semi–intensive) operations, promotes investment by government and military officials and by urban elites, and limits investment by less affluent and influential segments of Honduran society. The majority of farms under 70 hectares have been organized by urban investors with no previous experience in the industry and with insufficient capital to establish production. Many lack aeration equipment, technical assistance, and access to the transportation and marketing infrastructure. Such constraints are likely to result in the further concentration of shrimp farm holdings, as these marginal operations are unable to maintain production and repay loans (Casteñeda and Matamoros 1990).

The Effects on Coastal People and the Environment

Limited information from the region's communities that are most affected by the development of shrimp mariculture suggests that their household economies are organized much like those in more agriculturally oriented communities in Central America: they are remarkably flexible, dependent on remittances, and can shift among resources in response to changing market conditions and local resource availability (Stonich 1993). In these communities there appears to be considerable variability in socioeconomic differentiation as well; while approximately 25 percent of the households own the essential fishing gear (e.g., boats, nets, and motors), the remaining 75 percent of the households work as hired laborers for their more affluent neighbors. In this regard, ownership and control of the means of production (land, agricultural implements, boats, motors, nets, etc.) is similar to that found among agriculturalists, and is centered in the household. Although shared labor does occur, such sharing generally takes place among members of extended families. At the same time, the tradition of joint community action to acquire resources also exists.[9] For example, in 1992 twenty-one families from one coastal community engaged in a land occupation, and by August 1992 they had occupied and planted with corn, beans, watermelon, squash, melons, and manioc 40 *manzanas* from a total of 300 *manzanas* that they plan to clear and plant.[10]

Such groups are responding to the reduced availability of resources, that jeopardizes their livelihoods. Until the mid-1980s, when the construction of shrimp farms accelerated, the south's mangrove ecosystems provided a source of communal resources for families inhabiting the coastal zone. Many of these families were among those that had been displaced by the earlier expansion of cotton, sugar, and livestock in the region (Stonich 1993). Before the expansion of shrimp farms, the commercial and subsistence activities of these families had included salt production, fishing, hunting, shellfish collecting, tannin production, and fuelwood collecting. New land tenure arrangements, governed by the patterns of concessions and land prices, have been inclined to serve shrimp farms while placing traditional community land at jeopardy and fostering land speculation. Past events that removed small farmers from relatively good agricultural land, often by force and with the compliance of local authorities, have been repeated on the intertidal lands that have not been cleared. Wetlands, once open to public use for fishing, shellfish collecting, and the cutting of firewood and tanbark, are now being converted to private use.

Another source of friction has been the overlapping responsibilities of the various government agencies involved with different aspects of shrimp farming. Since the passage of *Decreto Ley* 968 in 1980, the Secretary of Culture and Tourism (SECTUR) has had jurisdiction over granting concessions because of its mandate to oversee state lands that border beaches and other tourist areas under that law. Before passage of this law, the National Agrarian Institute (INA) managed coastal land. At the same time, the Honduran Corporation for Forestry Development (COHDEFOR) had (until recently) the responsibility for protection and rational use of Honduras's forest resources and shared responsibility for the protection of the mangroves with the Department of Renewable Natural Resources (RENARE), which also has the right to supervise fishing and aquaculture within the country. The lack of unclouded demarcations of agency responsibilities has led to dissension and confrontation. Concessions are often granted without taking into consideration environmental suitability, the competence of the applicant, or even whether the current request overlaps with previous concessions (USAID/FEPROEXAAH 1989). Conflicts have arisen among the large foreign-owned operations, local medium-scale entrepreneurs, and peasant cooperatives over access, and between shrimp farmers and artisanal fishers.

At the same time, the extent to which the expanding shrimp industry can increase income for significant numbers of local people is questionable. According to a study of the labor force employed by twenty-nine shrimp farms, conducted in 1987, these operations created a total of 1,130 jobs (of which only 31 percent were permanent jobs), or 0.785 jobs per hectare of shrimp farm (less than one job per hectare). Of the total number of jobs, 93.5 percent were unskilled (manual laborers, assembly line workers, and security guards), 4.6 percent were administrative (administrative assistants and secretaries), and 1.9 percent were skilled (biologists and technical supervisors) (González 1987). In response to mounting public criticism of the industry, the National Association of Shrimp Farmers of Honduras (ANDAH), whose members tend to be owners and operators of larger farms, issued their own estimate of the number of jobs created: 1.5 jobs per hectare (ANDAH 1990).[11] Whichever estimate is used, the number of jobs created is insufficient in a region where close to 100,000 households are landless or land-poor and in which the estimated unemployment rate is more than 60 percent (ADAI 1987). In addition, while the majority of jobs are unskilled in all farm size categories, the overwhelming bulk of salaries is paid to skilled employees (Stonich 1991b). Local people are unlikely to hold such positions: many of the highest paying jobs in these categories are held by non-Hondurans, while lower-ranking administrative positions are filled by people from outside the local area (Stonich 1991b).

Diminished resources are also apparent in the recent declines in the number of finfish and shellfish captured since 1987, as reported by artisanal fishers. Destruction of habitats, blocking of estuaries, and rechanneling of rivers associated with the expansion of shrimp farms are among the factors that encourage ecological imbalances and destruction of other flora and fauna. The purported use of *Rotenone* by shrimp farmers to eliminate nondesirable species in ponds is also directly related to loss of stocks. Other interacting factors influencing this decline include increased sedimentation from erosion at higher elevations, a decade of drought, *El Niño* conditions, and the presence of pollutants from growing human populations and from the uncontrolled use of pesticides in the production of other export commodities (especially melons). Recent samples taken from Choluteca's shrimp farms showed relatively high levels of DDT in the young pond shrimp, which dropped below the maximum acceptable residue limits before the shrimp matured and were harvested. DDT or other pesticide residues have reportedly not yet exceeded international limits, but with each new pest outbreak in melons and each incremental increase in the volume and variety of pesticides applied, the risk that chemical contamination will affect the shrimp export industry grows (Murray 1991).

The Emergence of Grassroots Environmental Activism

The expansion of shrimp farms has taken place over the protests of individuals, communities, and environmental groups. The most significant of these is the Committee for the Defense and Development of the Flora and Fauna of the Gulf of Fonseca (CODDEFFAGOLF). Established in 1988, principally by artisanal fishers, to draw attention to the social and ecological problems associated with the expansion of the shrimp industry, the goals of this grassroots organization are to promote a balance among conservation, sustainable development, and social justice. CODDEFFAGOLF members have organized a sequence of protests: members have sent letters and proclamations to the President of Honduras, to the President of the Honduran Congress, and to the Commander in Chief of the Armed Forces; they have marched on the Honduran Congress with their demands; and they have physically blocked heavy earth-moving equipment. As an alternative to current destructive development practices in coastal areas, the group has proposed that the Honduran Congress create some variation of a national park or resource extraction reserve. The organization has generated a good deal of publicity and outside support for their goals among the Honduran public, the press, and international environmental groups.[12] Their growing renown was recognized at the United Nations Conference on Environment and Development (UNCED) in Rio de Janeiro in 1992, where they received a Global 500 award, which was accepted by one of their *campesino/* artisanal fisher members.

Issues of Equity and Sustainability

An examination of the social processes that have accompanied the expansion of shrimp mariculture raises significant questions regarding the distribution of costs and benefits and the extent to which this latest nontraditional commodity will enhance the income of people in the region. The earlier discussion of land tenure illustrates the unequal distribution of landholdings among shrimp farms. Habitat destruction and overcollection have exacerbated the problem of acquiring the naturally occurring postlarval and juvenile shrimp that are used to seed ponds, thereby raising costs, reducing profits, and forcing large companies to import seed stock from Miami and Panama or to attempt to establish their own nurseries (Castañeda and Matamoros 1990). A continued shortage in postlarval shrimp would most likely not only enhance the internationalization of the industry, through increased dependence on imported seed stock, but would also exacerbate the already unequal land distribution pattern. As has been the case, it is probable that only the large firms could afford to import stock in order to continue operation. This in turn might lead to increased expansion of these firms and to an even more skewed distribution of landholdings among shrimp farms.

Finally, the shrimp industry is vulnerable to the same international forces that affected other exports from the region. Factors such as reduced demand, falling prices, reduction in export quotas, and import restrictions due to higher quality control regulations, which had significant effects on the earlier cattle boom, are liable to influence the shrimp industry as well.

The Melon Boom

The most important of the other nontraditionals promoted in the south are melons—identified by USAID as being among the most competitive crops and able to generate the greatest employment and domestic income per hectare (USAID 1990:16). Eighty percent of Honduran melon production takes place in the southern region, principally in the departments of Choluteca and Valle (LACR 1989).

As part of its efforts to diversify production beyond bananas on the North Coast, the United Fruit Company began a melon export operation in the south in the mid-1970s through a subsidiary, *Productos Acuáticos y Terrestres, S.A.* (PATSA), which contracted with small and medium growers for rights to their melon crops. By the early 1980s, other producers began entering the market. Braced by stable prices and government incentives, growers aimed to expand the area under irrigation (more than 85 percent of melon fields are irrigated), improve the nutritional levels of their products, and expand distribution in U.S. markets during the winter season (LACR 1989). Between 1985 and 1989, the number of 15-kilo boxes of melons exported from Honduras rose from 600,000 to 2.4 million (valued at U.S. $8.5 million in 1989). By the 1989–90 growing season, 4,500 hectares were in melon cultivation, approximately half of which was being cultivated by small farmers (USAID 1990). Industry sources project an annual expansion rate of 600 to 700 hectares before reaching 6,000 hectares—the amount of land targeted for eventual conversion (Murray 1991).

The social and environmental consequences of the melon boom are as yet unclear, but several familiar predicaments are emerging. Boyer's budget studies of PATSA members in the late 1970s showed that the small and medium producers were barely breaking even (Janzen 1982). Murray's (1991) more recent study of the southern Honduran melon industry suggests that among the effects of expanded melon production are an escalation in the rate of land concentration (part of a shift toward greater control of melon production by large and transnational operations) and the concomitant elimination of independent small and medium producers. These results call into question the claim that this latest wave of nontraditionals will support small farmers and encourage equitable growth.

There is mounting evidence, as well, that melon production is exacerbating or recreating many of the same ecological disasters associated with the earlier production of cotton in the region. To control worsening pest problems, caused especially by aphid-born pest viruses, white flies, and leaf-miners, melon growers employ the same pest control strategy used in cotton—applying toxic chemicals on a calendar schedule. The effect has been the emergence of the familiar "pesticide treadmill" brought about as farmers escalate their use of chemical inputs in response to the increased resistance of pests to chemical strategies. Despite these efforts, melon farmers lost an estimated 10 percent of their melon crop to pests in the 1988–89 season and more than 50 percent during 1989–90 (Murray 1991). Although these losses were partially offset by high international prices, they call into question the long-term economic and ecological sustainability of current agricultural practices (Meckenstock et al. 1991). There is evidence, as well, that large-scale, especially transnational, producers are integrating anticipated escalating pest, economic, and social crises into their investment strategies. Murray (1991) reports a telling conversation in Choluteca with a spokesperson for the melon industry revealing what large growers in Honduras had learned from the earlier experience of their counterparts in Mexico (see, e.g., López 1990). Lamenting the possibility of a continued crisis in melons, the spokesperson remarked:

> Like the Mexicans, we may have to move if the problem isn't solved. We are currently looking into some disease-free zones in Nicaragua. If we have to, we can move these packing sheds and equipment in two days (cited in Murray 1991).

Other vital concerns are related to the social and environmental consequences of expanded irrigation for melon production. Water for drinking and bathing, as well as for irrigation, is becoming more scarce in the region. It is not known how the capture of larger amounts of water for irrigation will affect poor farmers' access to water, and their access to land suitable for irrigation. With no institutional regulation of well drilling, anyone with enough money to excavate a well can do so—and is doing so—in both rural and urban areas. For example, between 1987 and 1990, the regional office of the Central Bank was compelled to drill three wells, each one to a greater depth, because the previous well had gone dry (Castañeda and Matamoros 1990).

The lack of adequate surface water to meet increasing demands, several years of drought, extensive deforestation, and increased areas under irrigation appear to be depleting the coastal aquifer. Instead of contributing fresh water to the Gulf of Fonseca, the aquifer is now receiving Gulf water, and farmers with irrigation wells are encountering salty or brackish water for the first time. Reduction in the size of the coastal aquifer, coupled with contamination from leached pesticides, is already affecting water quality, human health, agriculture, and the shrimp industry (Castañeda and Matamoros 1990).

Conclusions

Promoted by bilateral and multilateral lending institutions, multinational corporations, and the state, the earlier expansion of nontraditional export commodities in southern Honduras, especially cotton and cattle, exacerbated inequalities and social differentiation. Capitalist agriculture was unable to create adequate employment, due to fluctuations in the world export market, land concentration, mechanization, and the replacement of crops by extensive livestock operations. The emergent semiproletarianized peasantry had access neither to sufficient land nor to employment opportunities to reduce poverty. The ensuing agrarian structure consisted of a complex network of classes and interest groups, each controlling different amounts of social power. Those groups that held stations of wealth and privilege were also inclined to have a greater voice in the realm of public policy; indeed, many of the main investors were important officials in the government and in the military. These more-dominant groups were least apt to experience the consequences of environmental destruction emanating from development efforts on a daily basis. In response to market forces, they shifted from investment in one export commodity to another with little regard for the environmental and social costs, and embraced emerging opportunities presented by the new nontraditional exports. One result has been the extension of catastrophic social and ecological processes to coastal zones. The privatization of state land attended by diminished available resources for communal use, escalating land values, rural displacement, and increased violence are not new phenomena in the south. Neither are massive deforestation and destruction of habitats emanating from the expansion of capitalist agriculture. A significant difference, however, between previous capitalist growth and the promotion of the shrimp industry is that the local people now being displaced comprise a semiproletarianized group that already had been partially dispossessed because of earlier capitalist expansion. The pattern in the growth of the shrimp industry has diminished the economic resource base of these families further. With record unemployment in the country, there is little hope that those displaced as a result of the expansion of shrimp farms can be absorbed into other sectors of the economy.

The promotion of the latest fashion of nontraditional exports in Central America was superimposed on an already-existing agrarian structure. Social and ecological problems similar to those associated with commodities promoted during the previous period have begun to emerge not only in southern Honduras but in other areas as well. Increasing evidence calls into question the assertion that small producers will benefit proportionally from the expansion of nontraditional exports (see e.g., Krueger 1989; Edelman 1990; Murray and Hoppin 1990; Murray 1991; Rosset 1991). Case studies from throughout the region reveal a number of factors that impinge on small farmers and the rural poor: rising rents and land values; access to credit and capital; government policies regarding prices and subsidies; inadequate access to technology and technical assistance; and unfavorable insertion into the market (Stonich et al. 1992).

Concerns over the environmental consequences of nontraditional exports are also emerging. Such issues include increasing incidence of pests and disease due to the introduction and intensive cultivation of crops; resulting ecological imbalances; crop, water, and air contamination stemming from improper use of pesticides and other chemicals; and the creation of "pesticide treadmills" (USAID 1989a; Murray and Hoppin 1990; 1992). The excessive and indiscriminate use of pesticides, a major problem for more than thirty years in cotton and other export crops, has become common in the production of nontraditional export crops throughout the region. This pattern of pesticide dependency and misuse poses a serious economic threat to the long-term development of the nontraditional sector, for growing numbers of produce shipments from the region have been confiscated and destroyed upon entry into the United States because of the presence of illegal pesticide residues (WHO 1989; Hansen 1990).

Also similar to earlier periods of "development" of the cotton and cattle industry, rural people from throughout Central America are attempting to modify, shape, and limit the powerful global forces impinging on their lives and livelihoods. The case of CODDEFFAGOLF is just one example of social mobilization and resistance that has emerged in response to the current round of debt-related, nontraditional export promotion.[13] The indications are that, as in previous export booms, the promotion of nontraditionals may accelerate social differentiation in rural areas, expel large numbers of small producers from their land, and lead to intensified human impoverishment, environmental destruction, and social instability in a region already rife with conflict. The continued refusal to recognize the potential social and environmental effects of the current export diversification model of development and the concomitant sacrifice of long-term sustainability to short-term advantage may cause ecosystems to exceed their limits and precipitate a permanent human and ecological crisis in the region.

NOTES

The research for this chapter was supported by the International Sorghum/Millet Collaborative Research Support Program through contract #AID/DSAN-G-0149, by the United States Information Agency through their Fulbright Senior Scholar Program with Honduras, by two University of Kentucky Summer Research Fellowships, by a University of California Academic Senate Research Support Grant, and by a University of California Pacific Rim Research Grant.

Special thanks are due Denise Stanley and Lyn Moreland for their preliminary fieldwork regarding the various effects of shrimp mariculture.

1. Analysis integrates qualitative and quantitative data collected between 1981 and 1990 from community and household surveys, statistical and documentary evidence, remote sensing and other geographic sources, and interviews with Central American farmers, fishermen, exporters, government officials, and representatives of national, bilateral, and multilateral development agencies. Further details on methodology can be found in Stonich 1992b, 1994.

2. The integration of human/cultural ecology and political economy has emerged as one of the major frameworks used to understand environmental destruction in the Third World (e.g., Blaikie 1985, 1988; Blaikie and Brookfield 1987; Little and Horowitz 1987; Redclift 1984; 1987). Political ecology as a theoretical approach, laid out by Blaikie and Brookfield (1987), includes the following essential elements: (1) Political ecology combines the concerns of ecology and political economy, thus integrating human and physical approaches to environmental destruction (1987:17); (2) Analysis follows a "chain of explanation" through different scales (levels of analysis), beginning with the decisions of local land managers, such as farmers, through the interrelations among local managers and other groups in society who affect local land management, and the roles of the state and world economy (1987:27); and (3) Because political economy insinuates analysis of structures external to local groups, considerable attention is focused on the ways international capitalism and the state affect natural resources and local people (Blaikie 1988). For examples of the use of political ecology in a variety of disciplines, see Schmink and Wood 1987; Bassett 1988; Sheridan 1988; Chapman 1989; Johnson 1989; and Saldanha 1990. For a comparison of contending paradigms used to explain environmental degradation in the Third World (Neo-Malthusian, classical-economic, dependency, political economic, and Marxist) see Stonich 1989, 1993.

3. Central American governments have been subsidized in their efforts to promote agricultural exports by many bilateral and multilateral international donors and lending institutions. During the 1980s, the region experienced a tremendous growth in so-called nonmilitary economic assistance from these agencies, creating dependencies that cushioned the effects of the worsening economies, as well as promoted nontraditional exports from the region. The majority of bilateral assistance came from the United States, as part of its efforts to stabilize regional governments and to appease populations in order to prevent successful popular uprisings, such as the Sandinista victory in Nicaragua in 1979. Although various U.S. organizations provided bilateral assistance to the region, the most significant was disbursed by USAID. Between 1946 and 1979, USAID dispensed a total of U.S. $1.9 billion to Central America; between 1980 and 1990 alone, it disbursed U.S. $7.5 billion (these and the following years and aid levels are calculated on the basis of government fiscal years and data from *USAID, Congressional Presentation*, various years.) During the second half of the decade, U.S. economic aid made up an average of 62.4 percent of total official development assistance (ODA) to the region (exclud-

ing Nicaragua, which did not receive financial assistance from the United States between 1984 and 1989).

According to USAID's new strategy for U.S. economic assistance to Central America in the 1990s, however, donor assistance from USAID will decline significantly, while aid from multilateral institutions is expected to be substantially larger than during the 1980s (USAID 1991b). As long as Central American countries continue to comply with the demands of the World Bank/IMF and IDB, disbursements of funds from those institutions should more than double during the 1991–96 period, compared to the previous six years. USAID also predicts a modest increase in other bilateral donor flows during the 1990s, which should allow a gradual decrease in U.S. assistance to the region. Under the scenario presented by USAID, assistance would decline from U.S. $810 million in 1990 (excluding U.S. $420 million in extraordinary assistance to Panama) to U.S. $430 million by 1996, for a total of U.S. $3.2 billion between 1991 and 1996. This would be a decline from the U.S. $6.2 billion, disbursed between 1984 and 1990. Over U.S. $1 billion in assistance during 1991–96 is expected to flow to Nicaragua. For the four countries included in the Central American Initiative during 1984–1989 (El Salvador, Guatemala, Honduras, and Costa Rica), the decline would be from U.S. $6 billion in 1984–1990 to U.S. $2 billion during 1991–1996 (USAID 1991b). Thus, through the 1990s, Central American countries will likely have to cope with reduced development assistance and enhanced indebtedness to the World Bank and the IMF.

4. The CBI grants duty-free treatment to all imports except textiles and apparel, leather goods, footwear, petroleum products, canned tuna, watches, and watch parts from twenty-four countries in the Caribbean and Central America. CBI beneficiaries are: Antigua and Barbuda, Aruba, Bahamas, Barbados, Belize, Costa Rica, Dominica, Dominican Republic, El Salvador, Grenada, Guatemala, Guyana, Haiti, Honduras, Jamaica, Montserrat, Netherlands Antilles, Nicaragua, Panama, Saint Christopher and Nevis, Saint Lucia, Saint Vincent and the Grenadines, Trinidad and Tobago, and the British Virgin Islands.

5. Recently elected Central American governments committed to market-based economic policies have responded to overwhelming pressure from international donor and lending institutions such as the U.S. Agency for International Development (USAID), the World Bank, the International Monetary Fund (IMF), and the Inter-American Development Bank (IDB) and have established severe economic adjustment programs whose objective is to restructure the economies through stabilization and growth. The principal measures implemented include: currency devaluations; liberalization of exchange rates that send the appropriate signals to importers and exporters; reduction of tariffs to neutralize the effects on sectoral development; decreases in public sector participation in the economy through privatization of productive activities and cuts in public expenditures (that give priority to external debt service); and utilization of tax systems to encourage exports and generate resources for expenditure priorities.

6. From 1983 through 1991, Belize supplied 4.2 percent, Panama 3 percent, El Salvador 2.3 percent, and Nicaragua 0.1 percent (computed from USDA 1992a, 1992b).

7. See Stonich (1991c) for a discussion of the growing dependence of Honduras on food imports and food aid, and nutritional consequences.

8. The recent expansion of shrimp farms is only one of the causes of the destruction of mangrove ecosystems. Approximately five hundred salt-making operations use mangrove wood as fuel in ovens used in the salt extraction industry: in 1988 these businesses used 50,000 m^3 of mangrove and other wood. In the same year, eight tanneries purchased 324 metric tons (approximately 1,500 trees) of mangrove wood for use in that industry. Finally, mangroves also provide an important source of domestic fuelwood, estimated to be 24,000 m^3 per year (SECPLAN/USAID 1989). Destruction of habitats, blocking of estuaries, and rechanneling of rivers associated with the above efforts also encourage ecological imbalances and destruction of other flora and fauna.

9. For discussions of peasant groups and resistance in the south, see White 1977 and Boyer 1982.

10. One manzana equals 0.69 hectares.

11. Other estimates of direct employment range from 1.2 jobs per hectare (including full-time farms and packing plants) to 5 jobs per hectare (on small farms) and from 3,000–8,900 full-time jobs.

12. In 1991, CODDEFFAGOLF received a grant of $100,000 from the Inter-American Foundation to finance and teach sustainable agricultural techniques to approximately 100 peasant families and to help 331 coastal families establish modern salt-evaporation and aquaculture ponds. CODDEFFAGOLF has also been successful in obtaining funding for conservation projects from the World Wildlife Fund.

13. For further discussion of CODDEFFAGOLF as both a social and environmental movement, see Stonich (1991b). For examples of recent peasant resistance to structural adjustment-related nontraditional export growth in Costa Rica, see Edelman 1990 and Rosset 1991.

REFERENCES

ADAI (Ateneo de la Agroindustria)

1987 Informe del Semenario-Lineamientos para un Mejor Aprovechamiento de la Ayuda Alimentaria. Doc. no. 42/87. Tegucigalpa, Honduras: ADAI. 26–29 de Octubre.

ANDAH (National Association of Shrimp Farmers of Honduras)

1990 Shrimp Cultivation: A Positive Support to the Development of Honduras. Choluteca, Honduras: ANDAH.

Banegas, Archaga; S. Volanda; Sonia M. Figueroa Cuellar; Yogena R. Paredes Nuñez; and Yoconda Colindres Zúñiga

1991 La Industria del Camarón en la Zona Sur de Honduras: Su Contribución al Mejoramiento Socioeconómico de la Zona." Master's thesis, Universidad Nacional Autónoma de Honduras, Tegucigalpa, Honduras.

Bassett, T. J.
1988 The Political Ecology of Peasant-herder Conflicts in the Northern Ivory Coast. Association of American Geographers, Annals 78 (3):453–472.

Blaikie, P.
1985 The Political Economy of Soil Erosion in Developing Countries. London: Longman.
1988 Land Degradation in Nepal. *In* Deforestation: Social Dynamics in Watersheds and Mountain Ecosystems, ed. J. Ives and D. C. Pitt, 132–158. London: Routledge.

Blaikie, P., and H. Brookfield
1987 Land Degradation and Society. London and New York: Methuen.

Boardman, R.
1986 Pesticides in World Agriculture. New York: St. Martin's Press.

Bottrell, D.
1983 Social Problems in Pest Management in the Tropics. Insect Science and Applications 4 (1–2):179–184.

Boyer, J.
1982 Agrarian Capitalism and Peasant Praxis in Southern Honduras. Ann Arbor, Mich.: University Microfilms.

Brockett, G.
1988 Land, Power, and Poverty: Agrarian Transformation and Political Conflict in Central America. Boston: Unwin Hyman.

Bruntland Commission (World Commission on Environment and Development)
1987 Our Common Future. Oxford: Oxford University Press.

Bull, D.
1982 A Growing Problem: Pesticides and the Third World Poor. Oxford: OXFAM.

Bulmer-Thomas, V.
1987 The Political Economy of Central America Since 1920. Cambridge: Cambridge University Press.

Buseo, J., C. Castaneda, F. Duarte, and M. Chavez
1987 Efectos de Plaguicidas en Honduras. Tegucigalpa: Universidad Nacional Autónoma de Honduras.

Castaneda C., Z. Matamoros
1990 Environmental Analysis for the Investment and Export Development Project. Tegucigalpa: USAID/Honduras.

Chapman, M. D.
1989 The Political Ecology of Fisheries Depletion in Amazonia. Environmental Conservation 16 (4):331–337.

CRIES (Comprehensive Resource Inventory and Evaluation System)
1984 Resource Assessment of the Choluteca Department. East Lansing, Mich.: Michigan State University and the U.S. Department of Agriculture.

DeWalt, B. R.
1983 The cattle are eating the forest. Bulletin of the Atomic Scientist 39:18–23.
DGECH (Direccion General de Estadistica y Censos)
1976 Censo Nacional Agropecuario 1974. Tegucigalpa, Honduras: Dirección General de Estadistica y Censos.
Durham, W.
1979 Scarcity and Survival in Central America: The Ecological Origins of the Soccer War. Stanford, Calif.: Stanford University Press.
ECLAC (Economic Commission for Latin America and the Caribbean)
1986 Central America: Bases for a reactivation and development policy. CEPAL Review 28 (April):11–48.
1987 Statistical Yearbook for Latin America and the Caribbean. New York: United Nations.
Edelman, M.
1990 When They Took the Muni: Political Culture and Anti-austerity Protest in Rural Northwestern Costa Rica. American Ethnologist 17 (4):736–757.
FAO-PY (Food and Agricultural Organization of the United Nations)
n.d. Production Yearbook. Rome: United Nations.
González, J.
1987 Situación de la Carcinocultura en la Costa Sur de Honduras. Honduras: RENARE.
Hansen, M.
1990 The First Three Years: Implementation of the World Bank Pesticide Guidelines 1985–88. Washington, D.C.: Consumers Union.
Heffernan, K.
1988 Problems and prospects of export diversification: Honduras. *In* Struggle against Dependency: Nontraditional Export Growth in Central America and the Caribbean, ed. E. Paus, 123–143. Boulder, Colo.: Westview.
Holdridge, L. R.
1962 Mapa Ecológico de Honduras. San José, Costa Rica: Tropical Science Center.
Howard Ballard, P.
1987 From Banana Republic to Cattle Republic: Agrarian Roots of the Crisis in Honduras. Ann Arbor, Mich.: University Microfilms.
ICAITI (Central American Institute of Investigation and Industrial Technology)
1977 An Environmental and Economic Study of the Consequences of Pesticide Use in Central American Cotton Production. Washington, D.C.: USAID.
Jarvis, L. S.
1986 Livestock Development in Latin America. Washington, D.C.: The World Bank.
Johnson, D. H.
1989 Political Ecology in the Upper Nile: Twentieth Century Expansion

of the Pastoral Common Economy. Journal of African History 30 (3):463–486.

Krueger, C.
1989 Development and Politics in Rural Guatemala. Development Anthropology Network 7 (1):1–6.

LACR (Latin American Commodities Report)
1989 Shrimp / Honduras. Latin American Commodities Report CR-89-07 (15 July):14.

Leonard, H. J.
1987 Natural Resources and Economic Development in Central America. New Brunswick, N.J.: Transaction Books.

Little, P., and M. Horowitz, eds.
1987 Lands at Risk in the Third World: Local Level Perspectives. Boulder, Colo.: Westview Press.

López, José Gabriel
1990 Agrarian Transformation and the Political, Ideological, and Cultural Responses from the Base: A Case Study from Western Mexico. Ph.D. dissertation, University of Texas, Austin.

Meckenstock, Dan, David Coddington, Juan Rosas, Harold van Es, Manjeet Chinman, and Manuel Murillo.
1991 Honduras Concept Paper: Towards a Sustainable Agriculture in Southern Honduras. Paper presented at the International Sorghum/Millet Collaborative Research Support Conference, 8–12 June, Corpus Christi, Tex.

Mejia, R.
1991 El Sector de Camarón en el Programa de Ajuste Estructural. Master's thesis, Universidad Nacional Autónoma de Honduras, Tegucigalpa, Honduras.

Murray, Douglas
1990 Export Agriculture, Ecological Disruption, and Social Inequality: Some Effects of Pesticides in Southern Honduras. Agriculture and Human Values 8 (4, Fall): 19–29.

Murray, Douglas, and Polly Hoppin
1990 Pesticides and Nontraditional Agriculture: A Coming Crisis for U.S. Development Policy in Latin America. *In* Texas Papers on Latin America, Paper no. 90-04. Austin Institute of Latin American Studies, University of Texas, Austin.
1992 Recurring contradictions in Agrarian Development: Pesticides and Caribbean Basin Nontraditional Agriculture. World Development 20 (4, April): 597–608.

Myers, N.
1981 The Hamburger Connection: How Central America's Forests Become North America's Hamburgers. Ambio 10 (1):3–8.

Nations, J., and D. Komer
1983 Rainforests and the Hamburger Society. Environment 25 (3):12–20.

Parsons, J.
1986 Forest to Pasture: Development or Destruction? Revista de Biología Tropical 24 (suppl. 1):121–138.

Parsons, K.
1975 Agrarian Reform in Southern Honduras. Land Tenure Center Research paper no. 67. Madison, Wis.: University of Wisconsin.

Paus, E., ed.
1988 Struggle Against Dependence: Nontraditional Export Growth in Central America and the Caribbean. Boulder, Colo.: Westview Press.

Perez-Brignoli, H.
1982 Growth and Crisis in the Central American Economies. Latin American Research Review 15 (3):365–398.

Redclift, M.
1984 Development and Environmental Crisis. New York: Methuen.
1987 Sustainable Development: Exploring the Contradictions. New York: Methuen.

Rosset, Peter M.
1991 Sustainability, Economies of Scale, and Social Instability: Achilles Heel of Non-traditional Export Agriculture? Agriculture and Human Values 8 (4, Fall):30–37.

Ruhl, J. M.
1985 The Honduran Agrarian Reform under Suazo Cordova. Inter-American Economic Affairs 39 (2):63–81.

Saldanha, I. M.
1990 The Political Ecology of Traditional Farming Practices in Thana Mararashtra, India. Journal of Peasant Studies 17 (3):433–443.

Schmink, M., and C. H. Wood
1987 The Political Ecology of Amazonia. *In* Lands at Risk in the Third World: Local Level Perspectives, ed. P. Little and M. Horowitz, 38–57. Boulder, Colo.: Westview Press.

SECPLAN/USAID
1989 Perfil Ambiental de Honduras 1989. Honduras: USAID.

Shane, D. R.
1986 Hoofprints in the Forest: Cattle Ranching and the Destruction of Latin America's Tropical Forest. Philadelphia: Institute for the Study of Human Issues.

Sheridan, T. E.
1988 Where the Dove Calls: The Political Ecology of a Peasant Corporate Community in Northwestern Mexico. Tucson, Ariz.: University of Arizona Press.

Stares, R.
1972 La Economía Compesina en la Zona Sur de Honduras: 1950–1970. Choluteca, Honduras: Unpublished report prepared for the Bishop of Choluteca.

Stonich, S. C.

1986 Development and Destruction: Interrelated Ecological, Socioeconomic, and Nutritional Change in Southern Honduras. Ph.D. dissertation, University of Kentucky, Lexington.

1989 Social Processes and Environmental Destruction: A Central American Case Study. Population and Development Review 15 (2): 269–296.

1991a Rural Families, Migration Incomes: Honduran Households in the World Economy. Journal of Latin American Studies 23 (1):131–161.

1991b The Promotion of Nontraditional Agricultural Exports in Honduras: Issues of Equity, Environment, and Natural Resource Management. Development and Change 22 (4):725–755.

1991c Lands and People in Peril: Ecological Transformations and Food Security in Honduras. *In* Harvest of Want: Food Security in Central America and Mexico, ed. A. Ferguson and S. Whiteford, Boulder, Colo.: Westview Press.

1992a Struggling with Honduran Poverty: The Environmental Consequences of Natural Resource Based Development and Rural Transformation. World Development 20 (3):385–399.

1992b Society and Land Degradation: Issues in Theory, Method, and Practice. *In* Anthropological Research: Process and Application, ed. J. Poggie, B. DeWalt, and W. Dressler, 137–157. Albany, N.Y.: SUNY Press.

1993 I Am Destroying the Land: The Political Ecology of Poverty and Environmental Destruction in Honduras. Boulder, Colo.: Westview Press.

n.d. Integrating Socioeconomic and Geographic Information Systems: A Methodology for Rural Development and Agricultural Policy Design. *In* The Anthropology of Human Behavior Through Geographic Information and Analysis, ed. M. Aldenderfer. N.Y.: Oxford University Press. Forthcoming.

Stonich, S. C., and B. R. DeWalt

1989 The Political Economy of Agricultural Growth and Rural Transformation in Honduras and Mexico. *In* Human Systems Ecology: Studies in the Integration of Political Economy, Adaptation, and Socionatural Regions, ed. S. Smith and E. Reeves, 202–230. Boulder, Colo.: Westview Press.

Stonich, S. C., D. Murray, and P. Rosset

1992 Enduring Crises: The Human and Environmental Consequences of Nontraditional Export Growth in Central America. Paper presented at the symposium, Local Impacts of Agricultural Restructuring. Ninety-first annual meeting of the American Anthropological Association, 2–7 December, San Francisco, Calif.

USAID (United States Agency for International Development)

1982 Honduras–Country Environmental Profile. McLean, Va.: JRB Associates.

1985 Environmental Assessment of the Small Scale Shrimp Farming Component of the USAID/Honduras Rural Technologies Project. Gainesville, Fla.: Tropical Research and Development, Inc.
1989a Agricultural Crop Diversification Export Project. Washington, D.C.: USAID.
1989b Strategic Considerations for the Agricultural Sector in Honduras. Tegucigalpa, Honduras: Office of Agriculture and Rural Development, USAID/Honduras.
1989c Environmental and Natural Resource Management in Central America: A Strategy for AID Assistance. Washington, D.C.: USAID.
1990 Agricultural Sector Strategy Paper. Tegucigalpa, Honduras: USAID/ Honduras.
1991a Congressional Presentation, FY 1992. Washington, DC: USAID.
1991b Economic Assistance Strategy for Central America, 1991 to 2000. Washington, D.C.: USAID.

USAID/FEPROEXAAH (United States Agency for International Development)
1989 Plan de Desarrollo del Camarón en Honduras. Tegucigalpa, Honduras: USAID/Honduras.

USDA (United States Department of Agriculture)
1992a Horticultural Products Review. Foreign Agricultural Service, Circular Series, FHORT 8–92, August. Washington, D.C.: USDA.
1992b Horticultural Products Review. Foreign Agricultural Service, Circular Series, FHORT 1-92, March. Washington, D.C.: USDA.

Weir, D., and M. Shapiro
1981 Circle of Poison: Pesticides and People in a Hungry World. San Francisco: Institute for Food and Development Policy.

White, L.
1987 Creating Opportunities for Change: Approaches to Managing Development Programs. Boulder, Colo.: Lynne Rienner.

White, R.
1977 Structural Factors in Rural Development: The Church and the Peasant in Honduras. Ph.D., dissertation, Cornell University, Ithaca, N.Y.

WHO (World Health Organization)
1989 Public Health Impact of Pesticides Used in Agriculture. Geneva: World Health Organization.

Williams, R.
1986 Export Agriculture and the Crisis in Central America. Chapel Hill: University of North Carolina Press.

World Bank
n.d. World Development Report. New York: Oxford University Press.

Chapter 3

Colonization, Development, and Deforestation in Petén, Northern Guatemala

Norman B. Schwartz

In the early 1960s, in the context of a growing agro-export economy, burgeoning population, and increasing land scarcity in southern highland Guatemala, the national government opened the northern lowlands to colonization and development. This was in part a substitute for the land reform that successive national regimes since 1954 had rejected as a way to ease the nation's social problems.

As a result, since the early 1960s the department of Petén has been settled and deforested at an accelerating pace. International donor agencies, technical experts, government officials, and intellectuals and poor farmers in Petén now find the rate and extent of deforestation alarming. Everyone blames everyone else for the situation. Government officials blame loggers; loggers and reformers blame government; local middle class authorities blame politicians in Guatemala City for caving in to popular demands for land; and they all blame the swidden farmers (*milperos*). The one point of agreement is that the scale of forest degradation is unwanted. Since everyone agrees on this, the question is: How did the unwanted situation come about? What historical, social, political, and economic processes led to what many think could become a disaster reminiscent of the collapse of Late Classic Maya society in Petén in the tenth century?

In general the answers are not novel. On the contrary, what is going on in Petén is dismally similar to what is occurring elsewhere in Central and South American (see for example DeWalt 1983; Heckadon Moreno 1983; Moran 1983; Partridge 1984). The details vary from place to place, but the processes and outcomes of new land settlement and forest degradation are roughly the same.

In what follows two major points will emerge. First, deforestation and inadequate reforestation are overdetermined. In Petén, relationships among small-scale farming, ranching, logging, uncontrolled colonization, and rapid economic and demographic growth drive deforestation and are in turn driven by macro-socioeconomic and political processes. This suggests that the impetus to deforestation may be stronger than the ability of anyone to retard it. Turning things around will require changes at the national and international as well as the local level. Recent government attempts to create protected forest zones inside Petén are laudable, but without changing conditions outside Petén they may be treating symptoms, not causes.

Second, although everyone would prefer to protect the forests, too many people benefit from deforestation to expect that calls for protection will avail. Tree removal is an outcome of rational (even if shortsighted) decisions. The immediate beneficiaries of forest degradation range from the poor *milperos* to wealthy loggers. To maintain the forests, beneficiaries will have to discover or be provided attractive alternatives, or the cost of cutting back the forests will have to be considerably raised.

Background

Petén, the northernmost department of Guatemala, comprises about 36,000 km^2, a third of the nation's total land area. From the Spanish conquest of central Petén in 1697 until the 1960s, the state and highland Guatemalans neglected the distant hinterland. As late as 1960 about 85 to 90 percent of Petén was covered with tropical dry forest north of and tropical wet forest south of Flores (the island capital of Petén) in the central lake district (fig. 1). A great savanna, located in south-central Petén, just south of Flores, covered about 8 percent of the department. During the colonial period (1697–1821), the cash economy of Petén was based on cattle and horse ranching, which declined during the 1840s. From the 1890s until the late 1960s, the basis of the economy was subsistence swidden agriculture and tree tapping—the extraction and export of chicle, a natural latex base for chewing gum. Until 1970 there was no year-round road access from Petén to the rest of Guatemala, and the internal road network was undeveloped. In 1964 the population was 26,720 people, of whom about 30 percent lived in Flores and the small towns of the central lake district.

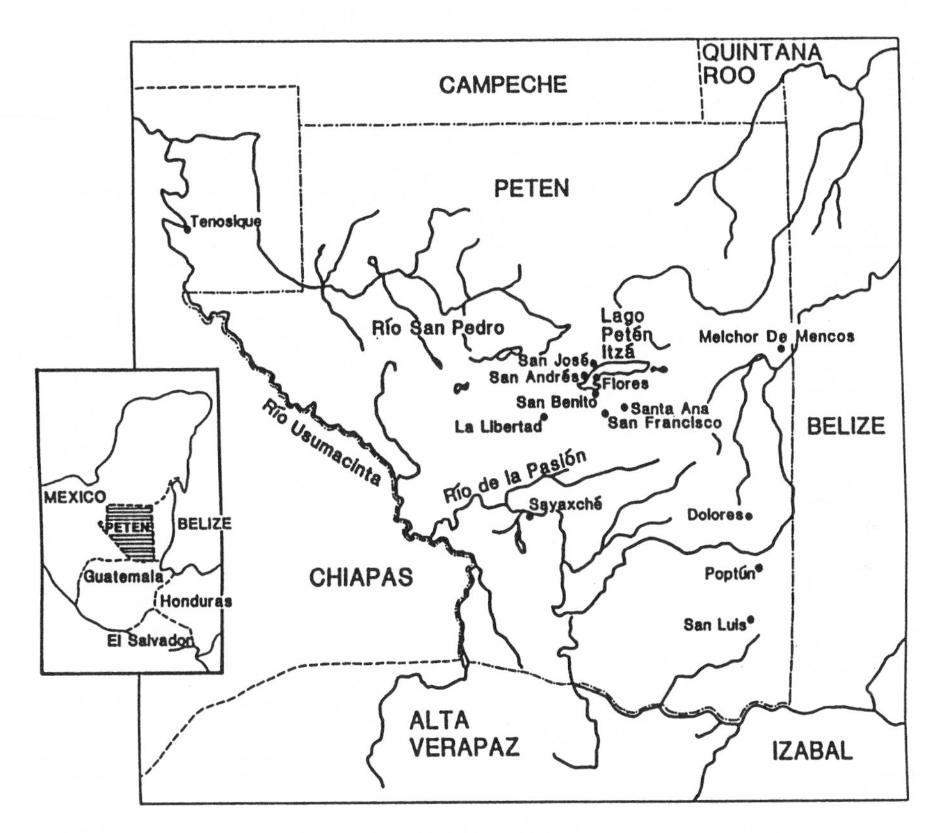

Fig. 1. El Petén

In 1959 the central government created FYDEP (Empresa Nacional de Fomento y Desarrollo de El Petén—National Enterprise for the Economic Development of Petén), an autonomous dependency of the national executive, to develop and colonize Petén. From 1959 to 1987, when the central government began to "liquidate" it, FYDEP had such extensive authority over Petén that critics scored it as a "state within a state." In 1959 about 98.5 percent of the department was in law a national estate. A mere 1.5 to 2.0 percent of the land was under private title.

According to its charge, FYDEP was (1) to administer and build an infrastructure to foment agricultural, industrial, and touristic development in Petén; (2) to administer and exploit the region's natural resources, except oil, for internal and overseas markets; (3) to sell land to landless countrymen for the production of basic grains (maize, beans, rice) and as a way to help reduce political demands for land reform in the south; (4) to "plant" settlers in cooperatives along the western borders with Mexico, to bar Mexican colonists from entering Petén and to prevent the Mexican government from constructing a hydroelectric facility near the border that might flood Guatemalan territory; and (5) to promote medium-scale capitalized cattle ranching in central and south-central Petén.

Technical studies indicated that about 33.3 percent of Petén should be dedicated to ranching, 17.6 percent to combinations of agriculture and ranching, 48.0 percent to managed forest exploitation, including controlled logging, and the remainder to urbanization and road construction (FAO/FYDEP 1970; Latinoconsult 1974). Ranchers could purchase from 225 to 675 hectares (later reduced to 225 hectares) and farmers 23 to 90 hectares from FYDEP. Because there were no immediate prospects for labor-intensive industries (other than logging) and because agricultural intensification could not be sustained, foreign experts recommended and FYDEP wanted gradual population growth—about 50,000 and certainly no more than 150,000 people in the foreseeable future.

Since the early 1960s there have been great changes in Petén, as even a brief review shows. In 1987 officials reported a population of 203,747 people for Petén (Prensa Libre 13 September 1987), but by my estimate the figure was closer to 300,000. Knowledgeable Peteneros say that 50 to 74 percent of the population are *sureños* (southerners, settlers from the south) and *Cobaneros* (Kekchí Indians from Alta Verapaz). Economic growth, too, has been impressive. There has been a mind-boggling expansion of commerce, government employment, construction, agriculture, ranching, logging, oil production, extraction of forest products (principally shate—decorative dwarf palms—and pimiento) for export, and tourism (now recovering from a decline caused by the civil war).

Although FYDEP has been charged with many faults, it did build a rudimentary infrastructure of roads, small bridges, a modern airport, grain storage facilities, health stations, schools, and so on. FYDEP, INDE (national electrification company), and local township governments have brought electricity to most of the central towns (*pueblos*, roughly equivalent to county seats) of Petén and to several larger villages subordinate to them. Much remains to be done: INDE's services are not reliable; isolated villages lack feeder roads and access to markets, schools, and much more; extension services are completely inadequate; and so on. Nonetheless, in 1970, when FYDEP opened an "all-weather"

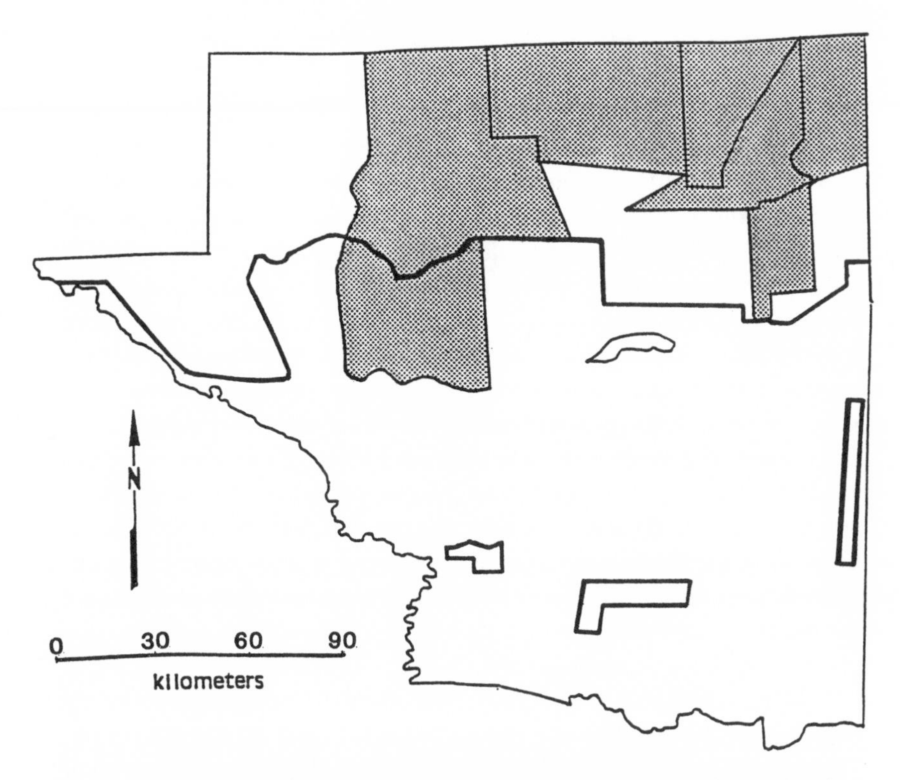

Fig. 2. Logging concessions and protected areas in Petén. Shaded area: logging concessions (approx.); thick lines: protected forests and parks

dirt road from central Petén to the central highlands, farmers in Petén began sending maize, beans, and rice to the south for the first time in modern history. In 1975 Petén was the leading producer of maize and beans in the nation.

Forests have been cleared as rapidly as population and economy have grown. By 1980 Peteneros were saying that 30 percent of Petén was deforested (also see Saa Vidal 1979). By 1990 forest inventories based on satellite imagery indicated that 40 percent of Petén was deforested and another 10 percent forest degraded (AHT/APESA 1992:Anexo 6 and 13). Moreover, despite regulations to the contrary, there are extensive agricultural fields in so-called forest reserves, north of latitude 17°10′. Around 1987, as FYDEP began winding down its operations, President Cerezo took steps to protect the remaining forests. Fig. 2 shows the approximate areas of Petén legally exempt from colonization.

National Context and Settlement of Petén

To understand how what no one wanted (so-called anarchic colonization and deforestation) but what no one halted came about, it is necessary to look briefly at modern Guatemalan history.

Since the 1870s relationships among the state, plantation owners, and rural labor largely defined the political economy of Guatemala. In the 1870s the state disallowed communal land tenure, which, in combination with coercive labor laws, assured plantations of adequate, cheap supplies of seasonal labor. Although the laws "did not register in a uniform way across the country" (Lovell 1988:104), many Indians in the Pacific piedmont, the central highlands, parts of the western highlands, and central Alta Verapaz did lose their land. By 1945 coercive labor laws were abolished; by then population growth, poverty, and land shortages among the peasantry made the laws unnecessary. Because Petén was not involved in the plantation economy; the government neglected it and its communication and transportation systems remained rudimentary. Nor, for reasons discussed elsewhere (Schwartz 1990), did chicle production generate backward internal economic linkages in Petén.

During the 1960s and the early 1970s the Guatemalan economy grew rapidly, led by increased production and export of beef and cotton. The commercial and financial middle classes and organized urban labor grew, but the benefits of economic growth went primarily and disproportionately to the traditional landed elite, bankers, industrialists, traders, professionals, and a growing list of multinational corporations operating in Guatemala. Plantations and ranches expanded, at the expense of displaced smallholders (Williams 1986).

As the elite extended its control over already very unevenly distributed land, the smallholders' situation deteriorated. For example, between 1964 and 1979 the number of the two smallest size farm groups (table 1), both of them unable to "fully employ the farm family and to produce enough income for family necessities" (Fletcher, et al. 1982:60), increased from 364,880 to 547,572. The average area of the two smallest size groups decreased from 0.38 to 0.24 hectares and from 2.17 to 1.53 hectares, respectively. The number of landless rural households nearly doubled, from 278,985 to 414,999, and the population grew much faster than urban and rural employment. Thus, many poor rural householders were forced to subdivide their small plots among family members, to migrate to Guatemala City and other cities, or to seek land in the north.

Increasing pressure on the land, high rates of population growth, limited urban-industrial employment, and growth-induced inflation increasingly worsened conditions for the dispossessed in the countryside and for redundant labor in the cities. The economic status of the poor deteriorated absolutely as well as relatively. By the late 1970s they were destitute (Adams 1970; Dixon 1987; Sexton 1985). Migration to the north became an attractive opportunity for some of the poor and a desperate last chance for others of them.

Multinationals and middle-class Guatemalans have also been drawn north. The multinationals are interested in the mineral and petroleum resources of the Northern Transversal (the northern part of Izabal, Alta Verapaz, Quiché, and Huehuetenango, immediately south of Petén), where they have driven up land prices, generating violent conflicts over land between newcomers and the indigenous Kekchí population. Many of the Kekchí have fled to Petén and Belize. In the process, military officers have acquired large tracts of land in the Transversal (Maloney 1981) and in Petén.

Many officers (and guerrilla leaders) come from upwardly mobile middle sectors, rather than from the traditional elite. Among the mili-

TABLE 1. Land Distribution in Guatemala, 1964 and 1979

Year	Size of Farms (in hectares)	Number	Percentage	Area (in hectares)	Percentage	Average Size (in hectares)
1964	<0.70	85,083	20.0	32,619.2	0.9	0.38
1979	<0.69	250,918	41.1	60,871.1	1.5	0.24
1964	0.70–6.99	279,797	67.0	607,855.6	17.7	2.17
1979	0.69–6.99	296,654	48.7	608,083.3	14.7	1.53
1964	6.99–45.13	43,656	10.0	648,900.2	18.8	14.86
1979	7.00–45.00	49,137	8.0	774,974.3	18.4	15.77
1964	45.13–902.51	8,420	2.0	1,258,545.2	36.6	149.47
1979	45.01–902.00	13,158	2.1	1,793,618.6	42.7	136.31
1964	>902.51	388	0.9	894,600.4	26.0	2,305.67
1979	>902.00	477	0.1	955,921.6	22.7	2,004.03

Farm area as a percentage of total national land area (10,888,900 hectares): 1964 = 31.6 percent and 1979 = 38.5 percent

Source: Fletcher, Graber, Merrill and Thorbecke 1970:59; Davis and Hodson 1982:45. Manz (1988:51) notes that between 1950–80, land per capita declined from 1.71 to 0.79 hectares. Between 1964–79, landless rural households increased from 278,985 to 414,999 (Early 1982:68–69).

Note: Farms smaller than 6.99 hectares accounted for 87.0 percent of all farms in 1964, and 89.8 percent in 1979. They held 18.6 percent of farm land in 1964, and 16.2 in 1979. Farms larger than 45.01 hectares accounted for 2.9 percent of all farms in 1964, and 2.2 percent in 1979; they held 62.6 percent of farm land in 1964, and 65.4 percent in 1979.

tary, whose salaries until recently have been relatively low, "ownership of land is given a particularly high priority" (Adams 1970:240), but opportunities to acquire land in the highlands have been preempted by traditional elites. Land is cheap and available in the Nothern Transversal and Petén, however, and some officers have obtained estates of up to 900 hectares at low cost along the fertile Río Mopán in eastern Petén (Gálvez 1982). Other middle-class Guatemalans—doctors, lawyers, architects, engineers, pilots from the national airline, university students, and so on (not to mention south coast cattle ranchers)—also wanted to acquire large tracts of relatively inexpensive land and even occasionally to actually settle in the far north (Schwartz 1987:174).

While people were moving north, the 1973 international oil crisis slowed Guatemala's economic expansion. After a brief recovery, the economy turned down for several reasons including the collapse of the Central American Common Market, an earthquake in 1976, a second oil crisis in 1978, growing inflation, and falling prices (beginning about 1975) for export commodities, particularly coffee and cotton. Between 1978 and 1982, the economy was further distorted by government corruption, civil war, and a 1982 unofficial currency devaluation. All this led several multinational corporations to withdraw from Guatemala and local capitalists to send their liquid assets abroad. The state was faced with a huge foreign debt, a recession, reduced revenues for social welfare, increasing socioeconomic inequality, growing rural and urban unemployment, and insurrection.

Since 1954 elite groups and the state have answered lower- and middle-class demands for economic and political reforms with repression. The military and right-wing paramilitary groups attacked not only insurgents but also middle-class reformers, moderate politicians, unionists, and rural organization leaders. In this process, the military took firm control of the state and large parts of the economy.

By the late 1970s an increasingly polarized polity and society had to cope with a deepening economic crisis. In 1982 world-wide recession threatened elite economic interests, aggravating an already explosive situation. "At precisely the time when elites were most threatened by the loss of land to the banks in the city, the peasants were moving onto lands in the countryside [and] the labor force was demanding higher wages" (Williams 1986:165).

By this time, guerrilla forces had established a wide base of support among rural Indians and Ladinos (united for the first time in the twentieth century) and controlled large sections of the western and northwestern highlands, where the Indians are in the majority. Unable to defeat the guerrillas, the army attacked their "civilian support base" (Black, Jamail, and Stoltz 1984:129). Between 1978 and 1985 the army waged a brutal war against the popular classes. At the time it seemed a "crime" to be young, or poor, or an Indian (Black, Jamail, and Stoltz 1984). The army also wanted to prevent Indians from building up their local economies and thus withdrawing labor from plantations (Lovell 1988:106). The Indians took the brunt of the war—thousands of them were killed and perhaps up to one million were displaced. By 1985 the guerrillas had lost the latest battle in an apparently endless war.

In 1985 the military permitted free elections, but, in an unstable alliance with the upper class, it retains its grip on the society and the state. For ideological and economic reasons, the officer corps and the socioeconomic elite continue to reject the reforms needed to resolve Guatemala's two most pressing problems—the role of Indians in the national society and the extremely unequal land distribution.

In the national context, it is no surprise that the central government has been unable, and probably has not wanted, to monitor closely or to finance adequately the settlement of Petén. From the 1970s to 1987 FYDEP's annual budget varied between Q3 and Q8 million. Depending on the year, the state contributed about 30 to 50 percent to the budget, and FYDEP contributed 70 to 50 percent, derived from its "private operations," such as land sales, fees, and taxes on the sale of chicle to foreign buyers, taxes on forest products (lumber, shate dwarf palms, and pimiento), FYDEP's own logging business, and various licenses. In some years income from logging was the single most important source of revenue, although in other years sale of land was. At least since the early 1970s, logging has accounted for between 16 and 54 percent of FYDEP's income. About half of the total annual revenues were usually spent on wages and administrative overhead. Thus in any given year FYDEP could invest no more than 50 percent of its income in development, much of it going to road and bridge construction and maintenance. In general, from 1974 to 1984 the cost for building a washboard dirt road in Petén varied from Q715/km on flat land to Q7,143 on rough upland terrain, the average being Q3,000 to Q4,000/km. Between 1974 and 1978 road maintenance costs averaged about Q390/km, and in 1984 about Q1,845/km (FYDEP 1974–85). These costs were high relative to FYDEP's budget, and it could barely construct more than 60 to 320 kilometers of road per year (and by 1990 there were said to be no more than 1,800 km of third class roads in Petén). As a result, hundreds of some one thousand small rural settlements in Petén lack feeder roads and are isolated during the rainy season. FYDEP's financial situation also helps explain why it lacked the trained personnel, equipment, and machinery to fulfill its mandate. A lack of continuity at the top administrative level leading to policy discontinuity, inability to withstand pressure from metropolitan politicians, and—according to FYDEP—the destruction of Q0.5 million worth of road-building equipment by guerrillas also limited FYDEP's effectiveness.

In addition, mismanagement, incompetence, favoritism, and venality have interfered with the effective use of revenues. Off the record, FYDEP officials—some of them dedicated, honest workers—concede all of this and more. But even if none of this had occurred, FYDEP's limited funds would have placed severe constraints on its capacity to colonize Petén in a sound way. Moreover it should be noted that in spite of many shortcomings, FYDEP accomplished more than is usually recognized. Nonetheless, FYDEP's meager resources and the national context within which the state decided to settle the north were not conducive to gradual, well-planned, and well-implemented colonization.

Farmers, Ranchers, and Loggers

Before turning to the settlers, FYDEP's land policies and rules are reviewed. Since the rules are apparently still in force, the present tense is used for part of the discussion.

FYDEP employs four criteria to set land prices: access to roads and markets, potential for commercial timber exploitation, soil quality, and water resources. A site may be paid for at once or in installments at low interest rates. After 20 years a title holder may sell his land, but until then the title cannot be transferred or sold without FYDEP's permission. Buyers who own more than 45 hectares, aside from a house site, outside Petén may not purchase land from FYDEP. FYDEP can rescind land that has not been worked for any six-month period; owners must keep 20 percent of their holding in forest; and buyers must reside in Petén. The regulations are supposed to prevent land speculation and to conserve the woods, but infractions are common. Initially, ranchers could purchase up to 675 hectares, but in the late 1970s the maximum was reduced to 225 hectares. FYDEP estimated that in most of Petén, a farm family (on average, six persons) needed 21 hectares to support itself—to crop 5–6 hectares for one to two years and to fallow the rest for three to five years (but many full-time farmers say they need 45 hectares to live above subsistence and to accumulate small cash reserves).

As noted above, foreign experts warned FYDEP that no more than 17.6 percent of Petén is "suitable" for farming, combined with small-scale ranching. Both the experts and FYDEP believed that *milperos* use "irrational" cropping techniques that imperil the forests. Thus, FYDEP was never eager to sell land to *milperos,* particularly poor Peteneros and Indian migrants. Instead, FYDEP preferred better "prepared" middle-class Ladino settlers. But, middle-level FYDEP managers claim that politicians in Guatemala City pressured them to sell land to thousands of spontaneous settlers from the south and to "plant" so-called cooperativists beside the Pasión and Usumacinta rivers, to prevent Mexicans from encroaching on Guatemalan soil. Some members of Petén's own upper and middle classes also favored large-scale colonization, arguing that demographic growth would increase commerce, supplies of labor, and so on, the benefits of which would eventually "trickle down" to everyone. FYDEP had to bow to the pressure, even though it was unprepared to cope with what one FYDEP official calls an "anarchic, uncontrolled invasion" after 1966.

Although a plurality of the "invading" *milperos* in the southeast, parts of the southwest, and along the southern border of Petén are Kekchí Indians from Alta Verapaz, elsewhere the majority of the settlers are Ladinos from the eastern highlands and southern coast of Guatemala (Centeno C. 1973; Fiedler 1983). In addition, hundreds of urban professionals, university students, and military officers, and some FYDEP officials have used their wealth and connections and loopholes in the rules to build estates that exceed 675 hectares (Góngora Z. 1984; Schwartz 1987). They own some of the best land in Petén, for example, in the fertile Mopán River valleys and plains of eastern Petén and in the southeastern Subín Valley area. Some do reside in Petén and directly work their land, some are absentee owners who hire estate managers, and others are absentee owners who hold unworked land for speculation.

Farm Households

There is a great deal of variation among farm households, but, after some preliminary remarks, the discussion will focus on two major types, located along the Pasión River and in central Petén. To avoid certain complications, most of the discussion refers to the years 1981–83, immediately prior to the worst years of the civil war in Petén, but the discussion will help explain why some poor farmers sympathized with the insurgents.

There are a fair number of Ladino settlers who, though not wealthy, have enough money and time to plan carefully the move to Petén and to select with care farm sites, usually along the better roads. Don Fulano, from Jutiapa, appears typical. After making several trips to Petén in 1970 and 1971, he bought 90 hectares near a village called Cascajal (psuedonym). In 1977 he bought a pickup truck and by 1985 had acquired another 45 hectares, about five kilometers from his first site. He felled hardwood trees on the second site (without FYDEP permission), sold the logs to a sawmill, and used the money to buy ten head of cattle. Don Fulano says he cleared the woods to put in pastures and also to secure his claim to the land. Don Fulano is meeting one of his major goals—saving enough money to educate his younger children for white-collar employment. He has done well in Petén because he came north with sufficient funds to buy good land near a road, has had good harvests, and has five healthy sons to work for him. In 1985 he felt that his family had a better life in Petén than they had or could have had in Jutiapa.

As described elsewhere (Schwartz 1987:169–71), until recently, relatively few poor native Peteneros purchased land from FYDEP, but members of the middle class did, and most of the traditional elite already owned estates. Some lower-sector Peteneros, taking advantage of the recent economic growth, have entered the middle class and bought land. Although their households resemble Don Fulano's in some ways, they differ in that they usually have more diverse sources of income. Their households typically include members who are forest collectors, labor contractors, artisans, merchants, professionals, and/or office workers, as well as farmers.

While hundreds of poor Ladino and Kekchí settlers have acquired good farm sites with relatively fertile soils near major roads, many more have had to accept less desirable sites, partly because before they came north so many better sites were already occupied by other farmers, ranchers, and land speculators. This said, discussion now turns to the Pasión and central Petén.

(a) Río Pasión

Many *milperos* located along the Río Pasión in the township of Sayaxché control 45 hectares of relatively good soil. Their major problem is lack of access to roads and markets, which forces them to sell to and buy at a disadvantage from a handful of middlemen who operate large, motorized canoes when the river is high. For example, in 1981 BANDESA (Banco de Desarrollo Agrícola, a government agricultural bank) calculated that in the township of Sayaxché the cost of producing 50 quintals of maize/hectare was Q309, but isolated riverine farmers could not sell their maize for more than Q6/quintal and often for as little as Q4/quintal, for a net loss of about Q9/hectare, and they were charged inflated prices for the few goods they bought from the middlemen. In 1988 riverine farmers were still selling maize for Q6/quintal to middlemen, who resold it for Q17–20. Two things partly offset this dismal picture: BANDESA calculated the opportunity cost of a day's labor at Q3, but farmers relied on household labor with lower opportunity costs, and they also sold other crops that had more value than maize, particularly beans. Nonetheless many, probably most, households were as poor in 1981 as they were in 1972, when almost all of them lived below the poverty line established by the government (Carranza F. 1973). Even before the civil war began and before rumors circulated that a proposed Mexican hydroelectric facility would flood their land, poverty forced many of these settlers to abandon their farms.

As for those who stayed, few can support themselves by relying exclusively on farming, although they are not short of land. To cope, they can crop more land or increase agricultural production without increasing plot size by using hybrid maize seed, but given the terms of trade, these are not prudent choices. They also can withdraw household labor from farming to seek seasonal wage labor, the major opportunities for which are in logging and in collecting nontimber forest products (NTFPs) such as shate, pimiento and chicle, or—usually for women—in retailing and domestic work in Sayaxché and other large towns. This choice swells the supply of labor and helps hold down wages, but on average smallholders who work off-farm part of the year do seem to be better off than those who do not. This is why articulate smallholders may decry deforestation but also fear attempts to curtail logging. Finally, they can seek land with superior road access, for example, near Cascajal, located alongside a major central-Petén road.

(b) Cascajal

In 1980 all but one male household head in the village was a settler (mean time in Petén was 9.4 years, in Cascajal 8.8 years). The majority were full-time farmers, cropping on average 6.7–7.0 hectares per year. At the time, they were reducing intercropping to grow more maize and beans because truckers from the highlands had little interest in other crops. In 1980 a Cascajal farmer could net Q52/hectare of maize and Q250/hectare of beans, which, while not grand sums, were much more than a farmer on the Río Pasión or in other isolated areas of Petén could earn. Under optimal conditions, a Cascajaleño who planted first-year plots half in maize and half in beans could net Q1,046 in cash and the equivalent of Q515 in "autoconsumption" (Carranza F. 1973:83)—enough to live above subsistence and perhaps enough to save some money (although consumer goods cost more in Petén than in the rest of the country).

But optimal conditions are not normal. Second-year plots may yield 35 percent less maize than first-year plots, and they demand more labor because of weed intrusion. Ill health, too, can reduce available labor. There also are fields that must be fallowed five or more years before they can be replanted. After some ten years of farming, many Cascajaleños found that the soil was less productive than it first was. Moreover, some *milperos* have had serious problems with crop pests, apparently for two major reasons. Reduced intercropping optimizes conditions for pest infestations, and farm plots are not widely separated by extensive areas of mature forest, which otherwise hinder the spread of insects (Wiseman 1978:93). Partly for these reasons, farmers say they need 45 rather than 21 hectares to avoid poverty.

In 1980 farmers were coping with their problems in several ways. Some household heads or their sons worked off-farm part of the year, but, at least in 1980, most were reluctant to spend months in deep forest felling trees or collecting NTFPs. Few had had experience with this sort of work, and in poorer households the first priority was to secure adequate supplies of food. Some intercropped more and planted less maize and beans, but because of the market this decreased cash income, making other choices more attractive. Others tried to intensify crop production. There are benign ways of doing so, for example, by spending more time protecting crops from animals, and less benign ways, for example, by shortening the fallow or by planting hybrid maize, which increases yield but has shortcomings (for example, owing to a loose husk, hybrid maize cannot be stored for long). For some, the easiest thing to do was to shorten the fallow, which meant they did not have to move far from the road (the effective market), but this added burdens to household labor, to combat weeds and so on, and reduced the time available for off-farm wage work. In addition, shortening the fallow degrades the soil.

Soils around Cascajal are made productive for short periods of time by burning forest biomass. Trees and bush are felled and dried cuttings burned to release nutrients stored in trees to upper soil layers. Firing a plot makes the soil more friable and kills insects that attack crops. In a plot that has been cropped for two years and fallowed for five to ten years, nutrients are reaccumulated by successional plants. "Leaf litter and dead vegetation return stored nutrients to the upper soil horizon, [releasing them] for building new plant tissue by decomposers in the soil." If the fallow is too short, however "heavy rains unrestricted by canopy or leaf litter, carry [organics and minerals] away . . . decreasing [soil] fertility" (Wiseman 1978:93). Weeding, by loosening the soil, accelerates the process. Moreover, without tree cover the rains can drive nutrients too deep into the soil for crop roots to reach them. Last, the soils around Cascajal are on well-drained slopes that are susceptible to erosion when cleared of protective tree cover (Sanders 1977). Thus, shortening the fallow can degrade the soil, and Cascajal farmers do not use chemical or animal fertilizers or compost that might otherwise compensate for loss of soil fertility. Yet, some farmers have had to shorten the fallow, although they know that this is a short-range solution to their problems.

To maintain adequate levels of crop production without shortening the fallow, some Cascajaleños have enlarged their holdings. But, because so much of the land near the road and around Cascajal was already occupied and because land prices had doubled since the 1970s, poorer farmers and latecomers, unable to acquire more land near the village, were moving deeper into primary forests and further from Cascajal, which increased the cost of transporting crops to the roadside and decreased farm income. Others sold their land (contrary to FYDEP's rules) to people like Don Fulano and left Cascajal. Finally, several men gave up traditional farming to grow marijuana, which since the late 1970s has become an important crop in parts of Petén.

As farmers (and ranchers) in Cascajal and nearby villages have spread out, hundreds of square kilometers of forest in the vicinity have been cleared or degraded. Over an extensive area, there are no dense protective stretches of woods between contiguous farms and ranches. As noted earlier, this increases crop pest infestations, especially when intercropping is reduced. Peteneros add that if a very large area of forest is cleared, seed-carrying birds and other animals will not cross it, thus retarding forest regrowth.

By the early 1980s, one important outcome of the above choices and processes was that, despite low population density (under 10 people/km^2) around Cascajal, farm sites with good soils and access to roads and to water were becoming scarce and concentrated in fewer hands. This led to several violent land disputes and to instances in which more powerful farmers ran poorer neighbors off the land. Overlapping land claims brought about by land survey errors, religious differences, and political instability added to the tensions in Cascajal. The worst event occurred in 1982 when several prosperous farmers reported to the army that certain poor farmers were providing insurgents with food. Some of the accused villagers were killed and some of them fled, which allowed those who denounced them to take over their land. The refugees moved to the comparative safety of the larger towns in central Petén, where the lucky ones secured work in the logging industry. The rest became poorly paid wage laborers or farmhands.

In general land concentration has been growing south of parallel 17°10′, where FYDEP sold land. Góngora Zetina (1984:125–31) reports that in 1980, 50 percent of landholders held 22 percent of this land, and 50 percent held the other 78 percent. By 1983 the first 50 percent held 13 percent of the land, the other 50 percent held 87 percent, and 5 percent of the owners held 56 percent of the land. Good sites are in short supply, affluent largeholders control much of the best land, and deforestation has become both an effect and a cause of land scarcity and concentration. Patterns of land tenure appear to be heading in the direction found in the central highlands of Guatemala.

Although no two households in Casjacal used the same mix of alternatives, *sensu composito* their actions had the same result. Within the constraints placed on them, individual householders made rational enough choices, but the the sum of the outcomes degrades the environment. This in turn forces poor householders to enter undisturbed forests, where the process may be repeated in ever larger, counterproductive cycles.

Ranchers

The Guatemalan beef industry took off in the 1960s. Between 1960 and 1980 the export value of beef went from $169,000 to $28,500,000 (Williams 1986:205–6). After 1980 exports declined, though domestic beef consumption has taken up some of the slack (Banco de Guatemala 1983). More to the point, in the 1960s and 1970s, when beef production was booming, FYDEP and foreign advisors believed that development of Petén would depend on ranching.

In 1970 FYDEP planned to sell 756,540 hectares, primarily to ranchers. In 1974 BANDESA opened its first branch office in Petén and by 1985 had seven offices there to promote farming and ranching. Simply by clearing land a rancher shows it is being "used" and thus is eligible for a BANDESA loan. However, affluent borrowers may not use bank loans for ranch or farm work. For example, one "big shot," who purchased 225 hectares for Q3,375, used a Q10,000 loan to invest in a hotel rather than in his ranch "because there is more profit in business." He seems not to have repaid the loan, and off the record a BANDESA agent says it is difficult to dun a "big shot." In any case, titled land can be used as collateral, encouraging wealthy people to acquire land they may not put to productive use.

Unfortunately, there are few reliable data on the number of ranches or cattle in Petén, but a former head of a ranchers' association estimates that in the early 1980s about 500 ranches functioned in Petén. The majority of them are located in the great savanna south of Flores, the area west of La Libertad, the upper Pasión near Alta Verapaz, upland valleys around Poptún in the southeast, and the Mopán River region in the far east. Farming is not particularly productive on the savanna, but it can be around Poptún and certainly is near the Río Mopán, where high-ranking military officers hold land.

As for the number of cattle in Petén, between 1957 and 1964 there were said to be 6,000 (Latinoconsult 1968). In 1977 FYDEP reported 21,000. By 1980 there were 150,000 (Zetina 1980:11), and by 1990 between 250,000 and 350,000. In 1980, by one estimate, half the ranches had herds of less than 100 cattle and half had herds varying from 100 to over 2,000 head.

In 1981 the Bank of Guatemala (1981) reported that many ranches in Petén had failed or were not profitable. The Bank says little about beef markets, but does say that inadequate road access, unacceptably low stocking rates (0.6 head/hectare), inexperienced management, use of methods that work on the south coast of Guatemala but not in Petén, and inadequate knowledge of which pasture grasses and which breeds of cattle are best adapted to Petén all contributed to the problem. In addition, water supplies are scarce in much of Petén during the dry season (roughly January through May). The bedrock in most of the region is porous limestone, which readily soaks up rainwater. Thus, streams, water holes, and wells in the great savanna and elsewhere are dry part of the year (Wiseman 1978:56). Moreover, in some places road construction and conversion of forest to pasture decrease the supply of water, prompting ranchers to relocate and to cut back more forest.

The Bank argues that for sound management, ranches of 225 hectares are best. Since 20 percent of a site must be kept in forest, 180 hectares can be in pasture. A ranch of this size can be managed by one family, with part-time help from another family. At 0.6 head of cattle/hectare, the ranch can stock 90 to 95 head, but the Bank recommends that stocking rates be increased to at least 1.1 head/hectare. According to the Bank, in 1980 a rancher had to invest Q84,000 to set up a 225-hectare ranch and to wait nine years before making a profit. Even if the rancher can wait that long, his per hectare returns may be less than returns from agriculture and cultivation of shate. Using the Bank's figures, a rancher who stocks about 1 head/hectare can net Q108/hectare; a farmer who cultivates maize and beans can net Q151/hectare (plus the value of food equivalents); and the owner of a shate plantation can net Q500/hectare. Although these admittedly rough figures, adjusted to 1987 prices, do not take into account discount rates and so on, the point is that, were the Bank correct, returns to ranching per unit of land would not be particularly attractive.

Nevertheless, ranching can have several advantages over other rural enterprises. Experienced ranchers can turn a profit in two to three years rather than the nine years the Bank reports and need not invest as much as the Bank reports to initiate a ranch. In parts of Petén stocking rates exceed Bank recommendations—from 1.9 head/hectare in the far southeast (Fisher 1974:348) up to 2.3 head/hectare on well-managed ranches 15 kilometers north of Flores, and up to 3.5 head/hectare on ranches using African grasses (*Brachiaria* spp.), recently imported through Brazil. In 1985 and 1987 a well-managed ranch with a permanent supply of surface water could net Q248 or more per hectare in some places, and a rancher does not need a large labor force, in contrast to a large-scale agriculturalist or shate contractor. The scale of an enterprise also must be considered. A ranch family with 180 hectares in pasture can net close to Q20,000/year, but a farm family must hire extra workers to crop 180 hectares, which will reduce their per hectare returns below the rancher's. Because of urban growth, tourism, and contraband trade with Mexico, demand for beef in Petén seems more stable than demand for shate and other NTFPs, which are notoriously vulnerable to market volatility (Heinzman and Reining 1988:52). Finally, land values went from less than Q14–27 (then worth $14–27) per hectare in 1974 to about Q270 (worth $54 because of currency devaluation) or more by 1987, so even a depleted ranch can be sold for a profit or used for collateral, logging, or speculative purposes. Ranching may not be good for Petén's forests, but it can be good for ranchers.

Although some ecology-minded ranchers are trying to maintain woodlots, in general, ranching contributes to deforestation, without compensatory generation of employment. For example, a rancher the Bank of Guatemala interviewed in 1981 gave up ranching in 1987, largely because of water-supply problems. He is fallowing (resting) his pastures, to replant them with maize and beans, and has used the proceeds from selling most of his herd to clear 25 hectares of virgin forest for combined farming and small-scale ranching. Some abandoned pastures may return to secondary forest in as little as ten years, but others—compacted, overexposed to the sun, and leached by hard rains—have to rest 25 years or more before they are productive again. In addition, there are localities like Cascajal where ranching worsens land scarcity.

Cascajal is surrounded on three sides by large ranches, several of which expanded into farming areas after their pastures failed. Farmers seeking new plots have had to leapfrog depleted and functioning ranches, that is, to farm far from the village (with all this implies for felling high forest), or have had to push out in what has become a crowded southerly direction. Growth and location of ranches have contributed to land scarcity in the vicinity. This is one reason some farmers left the village to clear forests elsewhere, and why others, unable to confront powerful ranchers, denounced their neighbors to the military. In Cascajal and other settlements, ranching has crowded farmers, thereby generating land disputes among them and accelerating deforestation.

Logging

Since the late 1970s Peteneros have been sounding alarms about uncontrolled logging, failures to reforest, venality in the industry, and so on. The same charges were made one hundred years ago (Valenzuela 1879). But, as noted earlier, so many people benefit from logging, which generates $7.7 million per year in revenue, that only the willfully optimistic would expect self-restraint in the industry.

In 1970 there were four logging concessions, located in northwestern, eastern, and southeastern Petén. In 1987 there were seven, covering much of the northern protected forest reserve area. There are about eleven licensed sawmills in Petén, plus some 20 clandestine sawmills that in collusion with officials cut and smuggle logs to Belize and Mexico. In 1985 FYDEP calculated that when fully operating the industry employs three to five thousand people, of whom several hundred are comparatively well-paid salaried employees (earning Q200 to Q500 in 1985). These figures do not include the many small-scale contraband loggers. Perhaps up to 10 percent of the households in Petén depend on logging for all or part of their income. Recall, too, that FYDEP also depended heavily on revenue from logging. Those who scored FYDEP for allowing too much logging and also for not building feeder roads were talking at cross-purposes. Funding for road construction has depended in part on logging.

FYDEP placed various restrictions on loggers (and from 1982 to 1984 suspended logging). Loggers were to plant ten trees for every tree felled, were not to use chain saws, were required to process much of their lumber in Petén, and so on. But FYDEP was unable to enforce its regulations. Aside from venality, it lacked the personnel, equipment, and funds to supervise the industry. Moreover, independent small-scale cutters and sawmill employees as well as powerful metropolitan interests object to attempts to curtail logging.

Just how many trees have been felled in recent years is anyone's guess. FYDEP (1974–1985) claimed that between 1977 and 1981 and during part of 1984 licensed loggers felled 44,103 mahogany, 16,603 cedar, and 7,333 other commercially valued trees—on average 11,339 trees/year, of which 66 percent were sent overseas. By 1987 the average was 14,792 trees/year, but this says nothing about underreporting by loggers, how many trees are left to rot when heavy rains make it impossible to haul them from the forest, or how many trees smugglers cut. For example, there are unofficial reports that in 1990 smugglers cut 20,000 trees in four months. And noncommercially valued trees are felled to get one of the valued but relatively widely dispersed trees. To make sure that a big mahogany tree falls freely to the ground, loggers fell 12 to 17 adjacent trees. This itself need not degrade the forests, but loggers cut down many other trees to open penetration roads to reach the mahogany trees. The roads enable farmers to move ever deeper into the woods, where they clear more forest. From the perspective of forest conservation, there is a negative symbiosis between loggers and settlers.

There is another, little studied, source of tree cutting—use of hardwoods for domestic fuel. A survey from central Petén showed that 96.7 percent of households relied on fuelwoods for cooking (author's field notes 1985). There is no reason to believe that the situation will change much in the near future. Firewood logs vary from about 3 to 20 cm in diameter and 0.6 to 2 meters in length (Wiseman 1978:109; author's Field Notes 1985), and

> firewood use rate per capita is fairly rapid. . . . A field abandoned two and one-half years near Santa Elena, Petén had softwood trees . . . up to 12.5 cm in diameter, but no hardwood species . . . over 2.5. Therefore, abandoned milpas . . . probably cannot support efficient wood harvest for well over five years. . . . This implies a considerable selective pressure on dense wood species with 2.5–12.5 cm [or 20 cm] boles. (Wiseman 1978:109)

Wiseman goes on to say,

> In an equilibrium model, three hectares of sustained yield land could supply the energy requirements of each person. I[n] an expansionist model . . . in which (forest) regeneration is not assumed, 0.6 hectare could support a person's fuel requirements. (Wiseman 1978: 109–110)

Rough though these estimates are, they do suggest that the demographic growth of recent years may be a greater threat to the forests than logging (for a comparable situation in pre-Columbian Copan, Honduras, see Abrams and Rue 1988).

By one report, farming is responsible for 29 percent of deforestation (Prensa Libre 29 August 1989:16), suggesting that logging, ranching, road construction, urbanization, and, above all, domestic fuelwood account for 71 percent. More important is that these activities feed on each other. Because thousands of farmers must work off-farm part of the year (and, reciprocally, wage workers must farm part of the year), licensed and contraband logging companies, NTFPs contractors, and merchants have access to abundant supplies of cheap labor. In turn, cheap labor helps makes logging (and other enterprises) profitable and encourages its expansion, and wages in the industry are attractive enough to assure it (as distinct from particular entrepreneurs) of popular support. Taxes on the industry are important sources of revenue for local and regional government, even though the tax rate is low and collection incomplete. So loggers continue to open penetration roads, making it possible for some farmers, pushed forward by other farmers and by ranchers, to enter formerly undisturbed forests so distant from markets that settlers must work off-farm as woodcutters (and harvesters of NTFPs). Ranchers convert forest to pasture without necessarily increasing herd size and without creating job opportunities that might otherwise withdraw labor from logging. True, almost everyone, including large-scale loggers and elite merchants, decries deforestation, but the elite welcome demographic growth, because, despite its implications for deforestation, it increases their wealth. In any case, the local elite cannot risk opposing conditions in the rest of the nation, which prompted the rapid colonization of Petén in the first instance.

Macrosocial conditions that led to the "anarchic" settlement of Petén worsened in the 1980s. The civil war of 1978–1984 intensified social and economic dislocations throughout Guatemala. The economy, which grew at less than 1.0 percent/year during the 1980s, cannot absorb a population that grew about 3.3 percent/year, to 9 million in 1989 from 6.5 in 1980. In 1980 some 0.4 million families lived in extreme poverty; in 1987, 1.1 million did, and 51.7 percent of the economically active population are still in the agricultural sector. For many, moving to Petén is a last chance to rescue themselves. Even before the 1980s thousands of poor rural families settled there, although FYDEP had not planned for and the state did not provide it with the resources needed for rational new-land settlement. Now, thousands of *milperos,* ranchers, loggers, and speculators not only compete for relatively easily exhausted land, but each also paves the way for or forces the other to clear more and more forest. The main beneficiaries of this overdetermined process are military, metropolitan, and regional elites, and elements of the middle class, but under current conditions it makes sense for wage workers and smallholders to fell trees. This may worsen things for the poor tomorrow, but first they must survive today. Deforestation serves the immediate interests of so many people and institutions that it will prove difficult to halt, short of turning the entire country around.

Nevertheless, there is a consensus that the remaining forests of Petén should be protected, and, some add, without harming lower-sector interests.

Forest Management and Human Welfare

Limitations of space prevent a discussion of the policies and programs government officials, international donor agencies, and others have put forward or are beginning to implement to manage the forests and (less often) to protect local interests in Petén (for example, see CONAP 1989; Garrett 1989; Góngora Z. 1984; Manger-Cats 1971; USAID Maya Biosphere Project 1990). Policies and programs range from possible revival of ancient Maya agrotechnologies, management and multiple use of forest resources, forest conservation, relocation of people already settled in protected forest zones, improved NTFP market access and harvesting techniques, and curtailed and controlled logging, to tourism development. The Cerezo government has placed agencies in Petén to guard protected forests, to educate and train extensionists and the general population in sustainable natural resource management, to do research on NTFPs, and so on. International donors plan to strengthen the agencies charged with the development of Petén and the management of its forests. Although the details cannot be entered here, several observations are in order.

Policies and programs now in place do not and cannot deal with the macrosocial conditions that threaten the environment, and they rarely offer immediate, concrete benefits to the lower sector. The structures of power that bar all this are precisely those that generated anarchic colonization of Petén in the first place. Current programs probably can slow the pace of resource depletion, but that may not be enough.

Partly in consequence, current policies may not be able to brake the expansion of swidden cultivation or overharvesting of NTFPs, with all this implies for forest depletion. As described earlier, given the terms of trade, many *milperos* must work off-farm part of the year, primarily in seasonal NTPF and logging industries, but their wages do not allow many of them to forgo farming altogether. True, over the last 30 years economic growth has led to a decline in the proportion of households deriving part or all of their income from farming, but even in the towns of Petén, most household heads make milpa. For example, in 1983–1985, 58.6 male household heads in one central Petén town made milpa, as did almost all heads of household in three central Petén villages (table 2). Because logging and chicle collection were curtailed during 1983–1985, no more than 31.0 percent of the men in the town were working off-farm in these industries, and only 19.9 percent of the men in the villages were. In the northern reserve, far from farm markets, 45 percent of villagers combined farming with NTFP collecting. Because overseas demands for NTFPs are labile, most householders perforce do and will continue to make milpa.

Markets for forest products such as NTFPs are notoriously unstable. For example, between 1940 and 1949 Petén exported on average 33,400 quintals of chicle per year. From 1980 to 1989 the average was 3,500 quintals per year, although in 1990, 10,000 quintals were exported. Between 1979 and 1987 Petén's share of the important U.S. shate market dropped from 37 to 17 percent, and Mexico's (a major competitor) went from 63 to 83 percent (Heinzman and Reining 1988:44). Overharvesting has less to do with collector and contractor ignorance than with the nature of boom-and-bust forest extractive economies, which encourage collectors to maximize harvests while they can. Telling hard-pressed collectors who must survive in the present that their harvesting methods negatively affect future income-producing opportunities is not useful. Of course there are ecologically sound ways to improve things, but they involve not only technological changes but also a restructuring of political-economic relationships at local and national levels (Heinzman and Reining 1988) not currently planned. Or, if the state interdicts overharvesting, collectors perforce will fell more trees for sawmills (licensed or not) or expand their milpas, though "continued expansion of milpa may be both economically inefficient and ecologically nonsustainable" (Heinzman and Reining 1988:64). As land in southern Petén becomes ever scarcer and concentrated in ever fewer hands, there will be that much more pressure on the northern forest reserves.[1] In addi-

TABLE 2. Occupation of Male Heads-of-Household in a Central Petén Town and Three Central Petén Villages, 1983–85

	Town		Villages	
Occupation	Number	Percentage	Number	Percentage
Farmer only	91	21.9	194	50.8
Farmer and farm hand	45	10.8	53	13.9
Farmer and logger	42	10.1	4	1.0
Farmer and forest collector	29	7.0	73	19.1
Farmer and other[a]	32	7.7	45	11.8
Farm hand only	4	1.0	4	1.0
Nonfarm labor (mostly in logging)	73	17.5	2	0.5
Forest collector only	27	6.5	1	0.3
Other (without farming)[a]	73	17.5	6	1.6
Total	416	100.0	382	100.0

Source: Author's surveys.

Note: Between 1983 and 1984, FYDEP tried, with some success, to suspend logging, thereby reducing the percentage of men who normally work in sawmills or as woodcutters during the dry season.

[a]"Other" includes ranchers, merchants, shopkeepers, NTFP contractors, civil servants (including teachers), office workers, pastors, curers, artisans (including construction workers), mechanics, truck drivers, motorized canoe operators, bee keepers, hunters, and fishermen. In 1985, there were seven or eight small-scale shate contractors in the villages.

tion, given the national shortage of land for a growing population in what is still basically an agrarian economy, and given the little chance for land reform, people now in Petén may be joined by more colonists from the highland seeking land and supplies of food in the lowlands.

Nor do current programs take sufficient account of Petén's regional power structure, which will ultimately qualify program implementation. Since this issue is discussed at length elsewhere (Schwartz 1990), here it may suffice to point out that the history of postconquest Petén and certainly the history of FYDEP demonstrate that policies and programs that fail to protect or at least prove neutral for elite interests will not protect the forests, much less the lower sectors.

But even on the rosy assumption that planners know how to assure positive outcomes, that they have sufficient funds to implement their plans, and that everyone cooperates, time itself is a problem. It takes time to implement complex policies (involving coordination among several agencies) designed to cope with the overdetermined deforestation process (USAID 1990). Until then, for rational enough reasons, people will continue to fell trees for firewood, to make milpas, to build roads, and so on. Forest degradation can proceed so rapidly that by the time programs to counter it are in place, there may not be much forest left to protect (cf. Heckadon Moreno 1985 on Panamá and Ross 1988 on the Lacandón rain forests of Chiapas, Mexico).

Conclusion

Deforestation is an outcome of rational decisions taken by individual households and entrepreneurs in the context of relationships among local groups and between them and supralocal processes. In Guatemala, for well over a century the state has served the interests of an agro-export plantation economy, and in place of popular support, the state has used repressive methods to control labor and divide ethnic groups to support that economy. The recent expansion of plantations and ranches, coinciding with rapid demographic growth, has led to increasingly unequal distribution of land. A long history of using repression in response to demands for reform has finally polarized the society and militarized the state. Insurgency, counterinsurgency, and a host of economic problems, many of them peculiar to dependent agro-export economies, have ground down the lower classes. All this pushes people out of the highlands and draws them to the lowlands, where land is or once was plentiful.

In the lowlands, the short-term interests of otherwise opposed social classes converge to generate a process of deforestation that, if it goes too far, may feed on itself, for deforestation reproduces many of the conditions that initially generate it. Within the context of Petén's relation to macrosocial structures and processes, relationships among *milperos*, wage workers, logging companies, ranchers, NTFP export houses, and the government overdetermine the pressure on the forests.

If this analysis is correct, then the actions needed to slow the pace of deforestation will have to match in scale and complexity the causes of it. Deforestation would palpably have to harm, rather than benefit as it now does, elite and middle-class interests and short-term lower-class interests. Short of using force, donor agencies, the state, and national elites will have to make sustained commitments to rational development in Petén, and probably to deep structural changes in national sociopolitical structures (implying some change in Guatemala's position in the world economy) to halt deforestation. Predictions are always risky, but the chances of all this happening in the near future seem slight.

Several things temper pessimism, however. There is general agreement in Petén, in Guatemala, and in donor agencies that the forests must be protected and restored. The government of Guatemala has lately (re)established forest reserve areas (and biospheres) and placed agencies in Petén to manage them. International donors have conditioned their aid to these agencies on ecological principles, and people in Petén at times do set aside short-term interests to safeguard their forest patrimony (Schwartz 1990: Chapter 6). Besides, gloomy assessments may be rejected on the ground that they abet passivity just when energetic action is needed. Planners would rather have specific, feasible solutions.

Thus there are ways, merely touched on in the last section, to protect the forests, not only by conservation but by wisely managed multiple-use exploitation that can benefit the lower sectors, but useful strategies will also have to promote, or at least not harm, the interests of powerful elite groups and the middle class. This has less to do with a preference for left-wing or right-wing politics and more with hard realities that must be recognized for exactly what they are and then must be changed. Conservatives and reformers, reactionaries and radicals, all may reject this conclusion, yet right now strategies that favor one class over another will not work. If the forests are to be saved, it will have to be for the sake of everyone in Petén and really for all Guatemalans.

NOTES

1. In the areas where settlement is legally permitted, there may no longer be much land available for distribution. The government is now trying to discover just what the land tenure situation is in Petén. At the risk of anticipating findings that may prove me wrong, as I understand the situation, it appears that FYDEP extended land titles to 4,593 people and made some sort of written agreement concerning land sites with another 2,511 people. An additional 22,435 people have begun the process of gaining legal possession of their sites, and officials estimate that about another 40,000 families are settled or squatting on the land (but some of them reside in Petén only during the maize-growing season). Thus some 69,539 families or individuals, the vast majority of them located in areas where settlement is legally permitted, hold some sort of possession of land. Of approximately 3.5 million hectares in Petén, about 800,000 hectares are in the Multiple Use Zone of the Maya Biosphere and another 807,000 hectares are in national parks, biotopes, and the like, which may not be exploited. This leaves about 1,893,000 hectares for most of the 69,539 families mentioned above, very few of whom hold less than 23 hectares. There still is land available in Petén, in part because an unknown number of settlers have abandoned their sites to return to the south, to move to towns in Petén, or to flee to Mexico (in part because of the civil war). Nonetheless, the above preliminary estimates suggest that Petén cannot absorb many more farm and/or ranch families than are already there, unless people invade protected forest areas.

Although there have been some changes (positive and negative) in Petén since 1989 when this paper was first written, processes driving deforestation have not changed, and so the conclusions reached here are still valid.

REFERENCES

Abrams, E. M., and D. J. Rue

1988 The Causes and Consequences of Deforestation among the Prehistoric Maya. Human Ecology 16:377–95.

Adams, R. N.

1970 Crucifixion by Power: Essays on Guatemalan National Social Structure, 1944–1966. Austin: University of Texas Press.

AHT (Agrar-und Hydrotechnik GMBH) y APESA (Asesoría Promoción Económica, SA)

1992 Inventario Forestal del Departamento de Petén. Anexo 6 and 13. Guatemala: AHT y APESA.

Banco de Guatemala

1981 Establecimiento de un Empresa Ganadera en el Petén. Informe Económico (April–June).

1983 Perspectivas de Producción y Exportación de Carne de Bovina. Informe Económico 30 (Jan.–March.):1–42.

BANDESA
1980 Memoria de Labores del Año de 1979. Santa Elena, Petén: Banco Nacional de Desarollo Agrícola.
Black, G. M., M. Jamail, and N. Stoltz C.
1984 Garrison Guatemala. London: Zed Books and the North American Congress on Latin America.
Carranza F., M. A.
1973 Cooperativas de El Petén: Producción y Cuentas Económicas. Guatemala: Universidad de San Carlos.
Centeno C., C. E.
1973 Cooperativas de El Petén: Situación Socioeconómica. Guatemala: Universidad de San Carlos.
CONAP (Consejo Nacional de Areas Protegidas)
1989 Estudio Técnico: La Reserva de la Biosfera Maya. Guatemala: CONAP.
Davis, S. H., and J. Hodson
1982 Witnesses to Political Violence in Guatemala. Boston: Oxfam America.
DeWalt, B. R.
1983 The Cattle Are Eating the Forest. Bulletin of the Atomic Scientists 39:18–23.
Dixon, W. J.
1987 Progress in the Provision of Basic Human Needs: Latin America, 1960–1980. Journal of Developing Areas 21:120–140.
Early, J. D.
1982 The Demographic Structure and Evolution of a Peasant System: The Guatemalan Population. Boca Raton: University Presses of Florida.
FAO/FYDEP
1970 Estudio de Preinversión Sobre Desarollo Forestal. 7 vols. Rome and Guatemala: FAO.
Fiedler, J. L.
1983 Commentary on Land Settlement in Petén: Response to Nancy Peckenham. Latin American Perspectives 10:120–123.
Fisher, G. R.
1974 Frontier Settlement Patterns in Northern Guatemala. Ph.D. dissertation, University of Florida, Gainesville.
Fletcher, L. B., E. Graber, W. C. Merrill, and E. Thornbecker
1970 Guatemala's Economic Development: The Role of Agriculture. Ames: Iowa State University Press.
FYDEP
1975 Petén Decada 70, Realizaciones FYDEP. Guatemala: Editorial Martí.
1974–85 Resumen de Labores. Santa Elena, Flores, Petén: FYDEP.
Gálvez, M.A.
1982 Arrasan Bosques Peteneros. Prensa Libre 26 July.

Garrett, W. E.
1989 La Ruta Maya. National Geographic 176:424–479.
Góngora Z., M. C.
1984 La Tenencia de la Tierra en el Departamento de El Petén y Su Legislación. Guatemala: Universidad de San Carlos.
Heckadon Moreno, S.
1983 Cuando Se Acaban los Montes: Los Campesinos Santenos y la Colonización de Tonosi. Panamá: Smithsonian Tropical Research Institute.
1985 La Ganadería Extensiva y la Deforéstación: Los Costos de Una Alternativa De desarrollo. *In* Agonia de la Naturaleza, ed. S. Heckadon Moreno and J. Espinosa González, 45–64. Panamá: Instituto de Investigación Agropecuaria de Panamá and Smithsonian Tropical Research Institute.
Heinzman, R., and C. Reining
1988 Sustained Rural Development: Extractive Forest Reserves in the Northern Petén of Guatemala. Guatemala: USAID/Guatemala.
Latinoconsult, S. A.
1968 Programa Ganadero de La Libertad: Introducción a la Crianza de Ganado de Carne en El Petén Central. Santa Elena, Petén: FYDEP.
1974 Estudio de Factibilidad de un Programa de Desarrollo de la Ganaderia Bovina en el Departamento del Petén. Guatemala: República de Guatemala.
Lovell, G. W.
1988 Resisting conquest: Development and the Guatemalan Indians. *In* Central America: Democracy, Development and Change, ed. J. M. Kirk and G. W. Schuyler, 101–107. New York: Praeger.
Manger-Cats, S.
1971 Tenencia de la Tierra y Desarrollo Socioeconómico del Sector Agrícola en Guatemala. 2d ed. Guatemala: Editorial Universitaria.
Manz, B.
1988 Refugees of a Hidden War: The Aftermath of Counterinsurgency in Guatemala. Albany: State University of New York Press.
Maloney, T. J.
1981 El Impacto Social del Esquema de Desarrollo de La Franja Tranversal del Norte Sobre los Maya-Kekchí en Guatemala. Estudios Sociales Centroamericanos 10:91–106.
Moran, E. L., ed.
1983 The Dilemma of Amazonian Development. Boulder, Colo.: Westview Press.
Partridge, W. L.
1984 The Humid Tropics Cattle Ranching Complex. Human Organization 43:165–75.

Ross, J.
1988 Mexico: Chimalapas Forest Falls to Loggers, Oil Pipelines, Poppy and Marijuana Fields. Latin American Press (Lima, Peru) 21 July: 5–6.

Saa Vidal, R.
1979 Mapa de Cobertura y Uso Actual de la Tierra. Guatemala: Secretaria General de Consejo Nacional de Planificación Económica.

Sanders, W. T.
1977 Environmental Heterogeneity and the Evolution of Lowland Maya Civilization. *In* The Origins of Maya Civilization, ed. R. E. W. Adams, 287–297. Albuquerque: University of New Mexico Press.

Schwartz, N. B.
1987 Colonization of Northern Guatemala: The Petén. Journal of Anthropological Research 43:163–183.
1990 Forest Society: A Social History of Petén, Guatemala. Philadelphia: University of Pennsylvania Press.

Sexton, J. D., ed. and trans.
1985 Campesino: The Diary of a Guatemalan Indian. Tucson: University of Arizona Press.

Simmons, C. S., J. M. Tárano T., and J. H. Pinto Z.
1959 Clasificación de los Suelos de la República de Guatemala. Guatemala: Ministerio de Educacíon Pública.

USAID (Agency for International Development)
1990 Maya Biosphere Project. Sanitized Version for Public Distribution. USAID/Guatemala Project No. 520–0395. Guatemala: USAID.

Valenzuela, S.
(1879) 1980 Informe Sobre el Departamento del Petén, Dirigido al Ministerio de Fomento. Reprinted in Revista Petén Itzá 21:19–56.

Williams, R. G.
1986 Export Agriculture and the Crisis in Central America. Chapel Hill: University of North Carolina Press.

Wiseman, F. M.
1978 Agricultural and Historical Ecology of the Lake Region of Petén, Guatemala. Ph.D. dissertation, University of Arizona, Tucson.

Zetina O., F. E.
1980 Petén Misterioso: 3 Alternatives. Revista Petén Itzá 21:11.

Part 2
South America

Chapter 4

Upland-Lowland Production Linkages and Land Degradation in Bolivia

Michael Painter

As anthropologists have become more explicitly concerned with the "political ecology" of production (e.g., Hjort 1982; Schmink and Wood 1987; Wolf 1972), it has become apparent that the social and economic relations that define the conditions of production determine how people use natural resources. Indeed, they define the critical natural resources to be controlled in a particular time and place, as well as who will have access to those resources and under what conditions. When access to critical natural resources is restricted so that people become relatively or absolutely poorer, for example, they also tend to become more extensive in their use of land, and to use land more destructively, deforesting larger areas in tropical lowlands and doing less to maintain soil fertility generally (Collins 1986, 1987). By the same token, when privileged interests are granted access to land on concessionary terms, they frequently are encouraged to be profligate resource users (Bakx 1987).

The same policies and practices that result in the wealthy receiving land on concessionary terms are responsible for the continuing impoverishment of poor people, because they institutionalize inequitable access to natural resources. Thus, the crucial issue underlying environmental destruction in Latin America is the gross inequity in access to resources. Conservation and development efforts that do not address this issue will be at best palliatives, and may have the long-term impact of intensifying land degradation and impoverishment (Painter 1990).

Studying and understanding the human relationships that drive processes of impoverishment and accumulation, and shape patterns of land use raises important methodological issues. The social relations that must be comprehended cover wide and environmentally heterogeneous geographic areas and do not necessarily involve face-to-face interaction.

As a result, researchers and planners frequently have difficulty identifying a unit of analysis or action that defines meaningful boundaries around their efforts and that continues to encompass critical areas and relationships.

An outstanding example of the difficulty, and the importance of overcoming it, may be found in the relationships that drive land degradation in the Andean nations of South America. Since pre-Hispanic times, sustaining production in the Andean region has depended on maintaining linkages between upland and lowland areas. The Inca state and its predecessors made control of vertically defined ecological zones a matter of state policy. In today's terms it was a national security issue, for the legitimacy of the state depended on its ability to guarantee access to the diverse goods needed to sustain human populations.

Conquest by Europeans marked the beginning of a continuing reorganization of production in response to the demands of international markets. The role of the state under the new order was to facilitate the extraction of commodities in demand at a particular time, and to reconcile the conflicting claims that interests tied to different productive activities would make on labor and natural resources. Responsibility for maintaining the vertical production linkages necessary for the reproduction of peasantry, on which the extractive structure depended, fell to local kin and community groups. At the same time, however, these groups were required to give up their best land and to provide the labor necessary for the functioning of the export economy.

Over the years the contradictory pressures exerted by the export economy have undercut the basis of vertically integrated production, and established downwardly spiraling cycles of economic underdevelopment and environmental destruction. The response to these cycles at all levels—rural households, commercial enterprises, state development policy, and development models promoted by international donor agencies—has been to seek out new land to bring into production. The social and economic problems afflicting areas already being utilized have remained unaddressed. The result has been to perpetuate the cycles of economic underdevelopment and environmental destruction, and to reproduce them in new areas.

The semiarid upland valleys of central Bolivia have suffered a long-term process of economic stagnation and underdevelopment that has resulted in severe land degradation, as a result of the relationships of unequal exchange established with the export mining economy in the

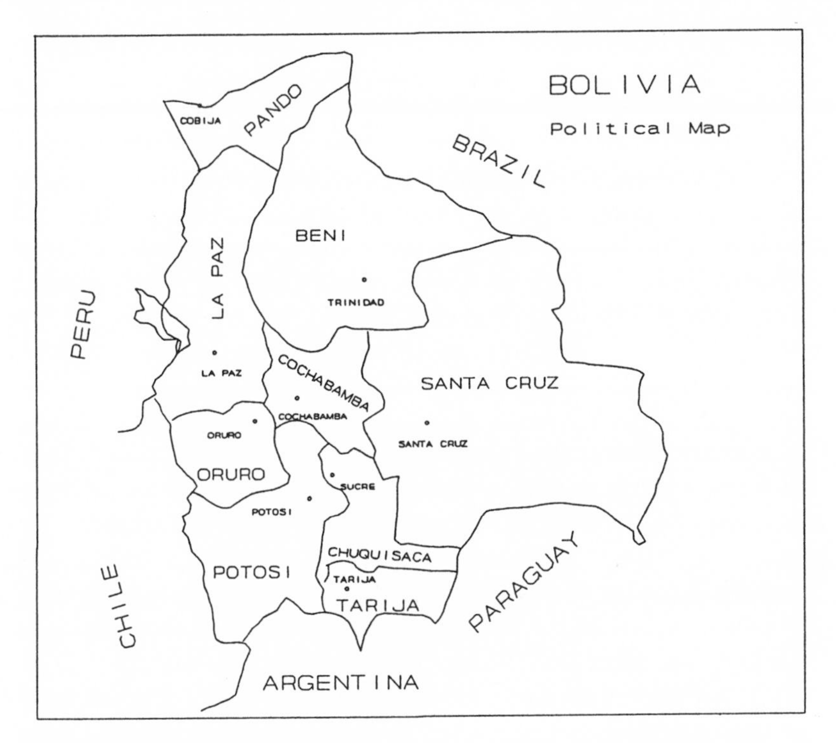

Fig. 1. Bolivia

colonial period. A series of events in the twentieth century have accelerated the process and intensified its impacts on the land. As a result, most of the rural population depends on activities off the farm for a major and growing portion of its income, a situation that itself intensifies a mutually reinforcing relationship between impoverishment and land degradation. One aspect of this situation is widespread involvement in agricultural settlement of lowland areas, including northern Santa Cruz department and the Chapare region of northeastern Cochabamba department (fig. 1).

The sharpening of impoverishment and land degradation in upland areas coincided with a boom in the international demand for cocaine. In the case of the Chapare, poor farmers, desperate to find a crop they could be reasonably sure of being able to sell, came together with entrepreneurs interested in responding to the opportunities offered by expanding cocaine consumption in the United States and Europe. As a result, smallholder settlement based on the production of coca leaf destined for transformation into cocaine expanded rapidly in the Chapare, beginning in the late 1970s and running through much of the 1980s.

Settlement of the Chapare raises a number of issues related to the impact of settlement in humid tropical areas. However, the crucial problem over the long term is that important elements of the social relations that resulted in the impoverishment of families in upland areas, and drove them to degrade farmland there, have been reproduced in the Chapare. Thus, it is probable that the cycles of impoverishment and land degradation will be repeated in the Chapare.

Underdevelopment and Land Degradation in Upland Valleys

Stagnation of the agricultural economy of Cochabamba's upland valleys was a development problem that dated from the colonial period of Bolivian history and periodically became a source of crisis. Agriculture has been closely tied to the mining industry, and the living conditions of rural dwellers have historically been affected by two major factors: (1) periodic droughts, which often caused crop failures in all areas not under irrigation and eliminated the possibility of earning income through the sale of agricultural products for most families; and (2) fluctuations in international ore prices, which affected the amount of food and labor exported from Cochabamba to the mining centers. The terms of trade between Cochabamba's agricultural economy and the mining centers tended to decline over time, so the export of workers to the mining centers became increasingly important. Of course, this intensified the region's vulnerability to the ebb and flow of the international mining economy, as Cochabamba supplied workers to the mines during periods of expansion, and reabsorbed people into the agricultural economy in periods of contraction (Dandler 1984; Harris and Albó 1986:36–54).

As the colonial economy evolved, the role of the agricultural areas was to supply mining and administrative centers with food and fiber. Smallholding farmers frequently undercut large estates in this enterprise, because they did not attach a value to their own labor, and could sell their produce at prices the large estates could not match if they were to be profitable. In fact, large estates could only count on making money in drought years, when smallholders were obliged to consume most of what they grew and had little left to sell. Because their landholdings were small and located in the least favorable areas for agriculture, however, many smallholders could not support themselves from farming, despite the fact that they dominated the markets for agricultural products in most years. As a result, smallholding farmers became heavily dependent on off-farm sources of income during the colonial period. During the eighteenth century in many areas of Cochabamba this stimulated the growth of craft industries, including textiles, pottery, and gunpowder in the Valle Alto, located to the southeast of the city of Cochabamba, and *charango* production further south in Aiquile, capital of Cochabamba's Campero province. Similarly, by the end of the eighteenth century, smallholders dominated the production and sale of bread in the city of Cochabamba, undercutting and eventually eliminating the established baker's guild (Larson 1988:202–205).

In the late eighteenth century, Intendent Francisco Viedma proposed the construction of a road linking the city of Cochabamba with the Chapare and promoted coca cultivation there as a way of infusing new life into the moribund regional economy. Although the road was not built because of opposition from other regional interests, exploitation of the tropical lowlands as a way to stimulate upland economic growth was an idea that anticipated modern development strategies by 200 years (Larson 1988:253–258).

During the twentieth century two events accelerated the impoverishment of rural Cochabamba: the completion of the railroad link between Cochabamba and Oruro in 1917, and the implementation of the agrarian reform in 1953. By connecting these cities with the Pacific Ocean ports (Mollendo and Arica), the rail network opened them as markets to such imported goods as textiles and housewares. The regional craft industries that had previously provided these goods were unable to compete with imported manufactured goods, so large numbers of people could no longer supplement agricultural production through work at home, and entered the migratory wage labor force.

The implementation of the agrarian reform in the high valleys of Cochabamba and neighboring departments also contributed to the inability of rural families to support themselves through agriculture. First, land redistribution was carried out more systematically in dryland areas than in irrigated bottomland, so the most productive land remained highly concentrated in the hands of few people. Second, with its focus on land redistribution, the agrarian reform did little to improve the productivity of agricultural labor in upland areas. While institutional mechanisms such as *sindicatos,* cooperatives, and producer associations have all attracted the interest of the national government and donor agencies, the investments in upland agriculture and associated improvements in market conditions, infrastructure, and the like have been small in relation to the investments made to support export production—mining in the highlands and valleys and commercial agriculture in Santa Cruz (e.g., Gill 1987; Frederick 1977; Heath 1969).[1] Thus, the development investment that accompanied the reform actually exacerbated the problem of unfavorable terms of trade with the export economy that the agricultural valleys of Cochabamba and neighboring departments had been experiencing for some time. Finally, as the formal barriers to access to schools, health clinics, and other services declined following the 1952 revolution, the subsistence requirements of rural dwellers rose, because they were obliged to contribute a significant portion of the costs of these services. These factors compelled people to enter the growing flow of population seeking income off the farm. By the late 1970s, 90 percent of the families in many parts of rural central and southern Bolivia earned more than half their income through off-farm sources (e.g., USAID/MACA 1981; Deere and Wasserstrom 1980).

Climatic adversity and changes in international commodity markets during the 1980s accelerated the processes of impoverishment and environmental destruction driving migration from upland areas. A severe drought that began in 1983 and lasted into the 1990s in some areas made life in their home areas impossible for many rural Bolivians. Northern Potosí was particularly hard hit, as may be seen by the numerous families from that area in the streets of Cochabamba and La Paz, begging, selling lemons, and seeking some way to survive in the cities. The drought pushed large numbers of smallholders throughout central and southern Bolivia "over the edge" in terms of their ability to earn a living through agriculture; thousands of families have left their homes permanently as a result, and thousands more have either begun to migrate seasonally or have increased the amount of time they spend away from home in search of employment.

Ironically, in some cases drought relief efforts have themselves increased the pressure on people to migrate. In Mizque, for example, many farmers received agricultural credit under the *Crédito Agropecuario de Emergencia* (CAE) program, sponsored by USAID/Bolivia, with participation of P.L. 480 and some *Instituciones Crediticias Intermediarias* (ICIs). The credit program was intended to speed the recovery of rural families affected by the drought by easing access to seeds, inputs, and improvements for their land. Unfortunately, in many areas such as Mizque, the drought lasted longer than the CAE program. Farmers had to repay their loans even though the drought had not ended and harvests were still below normal. In order to repay, many families sold livestock, farm implements, and even tried to sell or give their land to the financial institution, efforts that were generally inadequate. When USAID/Bolivia sought money for the credit component of its project to provide farmers growing coca leaf, it began to pressure the financial institutions involved in the CAE program to collect the outstanding loans and return the money to P.L. 480. Naturally, the financial institutions increased the pressure on farmers, who responded to the increased cash demand by migrating to the Chapare to earn money. Sometimes, the loan program was responsible for the migration of families who had not previously migrated (Cuba 1989).

In 1985 a second financial disaster struck the poorest sectors of Bolivian society, as international tin prices collapsed when the London Metal Exchange terminated trading. The result of an unfavorable shift in exchange rates and the accumulated debts of the International Tin Council, the event brought ruin to Bolivia's principal legal export industry. Between August 1985 and August 1986 some 27,000 mine workers lost their jobs. The *Banco Central de Bolivia* estimated the unemployment rate to be 20 percent by the end of 1985, largely because of the layoff of mineworkers, and, according to the *Central Obrera Boliviana* (COB), the figure approached 30 percent by the end of 1986 (Crabtree et al. 1987:20). The impact of the mining collapse on families not directly employed by the mining industry—but dependent on it—has never been measured. Many families migrated to urban areas, particularly Cochabamba and La Paz, and from Cochabamba, many went to the Chapare when they were unable to find work in the city. Others went directly to tropical lowland settlement areas in Pando and Beni departments, as well as to the Chapare region.[2] For many, the Chapare was the destination of choice because coca leaf production offered an immediate source of wage labor.

These events took place in the context of general financial collapse, a collapse closely related to Bolivia's inability to continue payments on its substantial foreign debt, most of which was incurred during the 1970s under the Banzer regime. In addition, in 1983 the Siles government embarked on an ill-conceived attempt to unlink the exchange rate of the Bolivian peso from the U.S. dollar. The effects of this *desdolarización* on the already weakened Bolivian economy included a disastrous annual inflation rate that exceeded 14,000 percent at its peak in 1984. Even middle-class salaries evaporated, and only those with access to dollars enjoyed any protection. Lacking alternatives, many people turned to participation in producing or processing coca leaf.

In response to the hyperinflation, the Bolivian government of President Paz Estenssoro implemented an economic stabilization program that resembled in many respects the one implemented by the Siles government in 1957 (Dunkerley and Morales 1986; Urioste 1989). A major impact in rural areas was accelerated economic decline. In Aiquile and Mizque provinces of Cochabamba department, a substantial portion of whose rural population depends on wage labor in the Chapare, overall potato prices declined by 9 percent, while the price of the chemical inputs required for commercial potato production increased by 100 percent, during 1988–89. The price of seed potatoes, which is the major cash crop for smallholding families in many upland areas of these and neighboring provinces, declined by 20 percent during the same period. In the Valle Alto of Cochabamba, which has long been a major source of Chapare settlers and laborers, the cost paid by farmers to transport agricultural produce increased by 33 percent, input costs went up by 15 percent, and the cost of dollars (needed to secure the inputs) increased by 160 percent between 1987 and 1989. Farmer earnings associated with the production of a high-value potato variety declined from $1000 per hectare to $650 per hectare (Healy 1991).

Such economic hardship manifests itself in several areas of rural family land use. Two of the most apparent are deforestation resulting from the encroachment of individually held farmland onto forest and brushlands that previously had belonged to the community, and the growth in importance of livestock production. Various factors drive deforestation. For example, the decline in rural incomes has resulted in a growth in the number of people involved in such longstanding rural industries of the area as the brewing of corn beer, or *chicha*, and brickmaking, industries that consume large amounts of firewood (Pérez 1991a).

Land scarcity also contributes to deforestation. A recent study found that 63 percent of 360 household heads interviewed in the provinces of Campero and Mizque in Cochabamba department acknowledged having cleared forest in order to gain access to arable land (CEFOIN 1990:57). Land scarcity is commonly attributed to population growth, but this explanation is unconvincing because the population in the area has remained relatively stable, with growth being offset by outmigration. Land scarcity seems to be more a product of declining productivity, resulting from the labor scarcity caused by drought and by reliance on off-farm employment. Families require more land to grow the same amount of produce, and parents hold on to land longer before allowing offspring to inherit. This results in increasingly scarce farmland for newly formed households, and generates pressure on communities to allocate forest and grazing land that had been held in common for use as cropland by individual families (Gisbert and Quitón 1991; Gisbert et al. 1994).

Finally, labor scarcity generated by high rates of outmigration contributes directly to deforestation. Rural women in dryland areas of Campero and Mizque provinces spend two or three hours a day gathering water, as much as four hours a day in food preparation, and two or three hours a day gathering firewood. These domestic chores are managed together with agricultural production and livestock management. With such demands on their time, strategies that existed in the past to manage fuelwood resources and spread demand over a larger area disappear, and women forage for wood as near to home as possible. The result is an ever-widening deforested area around each house (Gisbert and Quitón 1991; Gisbert et al. 1994).

Environmental problems resulting from livestock management practices are interrelated with those of deforestation. The culturally defined norm for livestock ownership is that large livestock, primarily cattle, are owned by men, while small animals, among which the goat is most important, are owned by women. Goats are usually allowed to roam freely in the vicinity of the home, and are the most numerous and visible livestock in the area. Considerable environmental damage is attributed to them, resulting from the impact of their browsing on trees and shrubs in the semiarid environment and the paths they create on the steep hillsides.

In the face of widespread male outmigration, women also assume full responsibility for the care of cattle. In the context of the other chores that fall to them, care of the cattle is a burdensome addition. This is particularly the case in those areas where cattle are kept in lands, called *monte*, that belong to the community or the local peasant union (*sindicato*), to which families have use rights, which are often located at some distance from where the families who use them live.[3]

The problems of impoverishment and land degradation come together around livestock in several ways. Because of the labor scarcity resulting from the high rates of male outmigration, livestock management practices deteriorate. In the case of cattle, labor scarcity makes it difficult for women to make long walks to *monte* areas as often as they should. Salt licks are not maintained; the condition of the cattle is not closely monitored; and they are not always moved as soon as they should be when grass and forage begin to deteriorate in a particular area. Similarly, goats are allowed to roam relatively freely in the vicinity of the home, with little control over their movements. The destruction resulting from their unmanaged browsing compounds the destruction associated with women's woodgathering, severely degrading the immediate surroundings of family compounds. Also, the uncontrolled movement, combined with the labor scarcity, means that much of the manure produced by the goats goes unutilized, even though it is one of the resources for which goats are valued, and crops are denied an important source of fertilizer.

Two factors compound the deterioration of livestock management. One is the encroachment of agriculture onto land that had been reserved as community forest or pasture and forage areas. The conversion of land to a use that had been regarded as inappropriate, and the greater intensity of use of the remaining pasturage by animals are both sources of degradation. In addition, the number of livestock making demands on the land is growing as an outcome of impoverishment and growing dependence on off-farm income. The scanty statistical data available on the area show a steady increase in the number of livestock through the 1970s (e.g., CIDRE 1987:211). Farmer responses in interviews state that this trend continued and accelerated during the 1980s, because of several factors.[4] In the absence of rural financial institutions, livestock is the only form of savings available to families when cash earnings are not totally consumed in meeting immediate needs. Animals and animal products assume greater importance in the context of the frequent crop failures experienced as a result of the prolonged drought of the 1980s.

The combination of more animals being maintained on smaller pasture and forage areas has begun to generate new types of human pressure on the environment. For example, in communities where livestock are maintained in *monte* areas, farmers report a marked increase in the last two to three years of attacks on farm animals by the Andean puma. This predator had succeeded in living relatively invisibly in the *monte* areas for many years, but encroachment by people and livestock has reduced their range and destroyed habitats on which their normal prey depends. Already endangered, pumas are now coming under a double pressure as the incidents of livestock killing have piqued farmer interest in hunting the cats down and killing them in order to protect their animals.

Patterns of Migration and Settlement

Settlers in the Chapare

Migration is a response to the impoverishment and land degradation afflicting such upland areas as Campero and Mizque provinces, and the labor scarcity provoked by migration perpetuates the mutually reinforcing relationship between the two processes. The most frequent rural destination for migrants from much of central Bolivia during the 1980s and early 1990s has been the Chapare (Blanes 1984; Flores and Blanes 1984; Painter and Bedoya 1991; Pérez 1991b). This importance was heightened by the emergence of the Chapare as a major source of coca leaf for cocaine production. However, migration out of central Bolivia began well before the international boom in cocaine demand, and most migrants continue to seek work in areas other than the Chapare. Bolivian cities, particularly La Paz, Cochabamba, and Santa Cruz, have been the largest overall recipients of people from rural areas (Pérez 1991b). Similarly, migration to provide agricultural labor in Argentina (e.g., Whiteford 1981), unskilled or semiskilled labor in Buenos Aires (e.g., Balán and Dandler 1986), and agricultural labor and settlements in Santa Cruz (e.g., Hess 1980; Stearman 1976, 1985; Painter 1987; Painter and Partridge 1989; Pérez 1987) has been prominent over time in different parts of central Bolivia. The importance of one or another migratory destination rises and falls with the opportunities that different areas offer for securing land and/or work. Many families have maintained contacts in different areas, and respond rapidly to changes in the opportunities and costs associated with each (e.g., Balán and Dandler 1986; Carafa et al. 1987).

One begins to appreciate the resource scarcity that motivated families to settle in the Chapare by examining the situation regarding the agricultural land they left behind. As part of the research being described, the *Centro de Estudios de la Realidad Económica y Social* (CERES), a private Bolivian social science research institution, interviewed 194 Chapare farmers regarding their settlement experiences and the agricultural production systems.[5] Of the 194 farmers interviewed, 176 responded to questions about why they had moved to the Chapare. Of these, 143, or 83 percent, said they had migrated because of a lack of economic opportunity, in the form of inadequate land or income, at home. Of the 194 farmers interviewed, 187 provided information on their landholdings in their communities of origin. Of these 187, 111, or 59 percent, reported that they did not own land in their home communities, meaning that they had been landless before they migrated, or that they had given up their claims to land in their home communities after settling in the Chapare. Of the remaining 76 farmers who continued to own land in their home communities, 48, or 64 percent, of them owned one hectare or less, while 92 percent owned three hectares of land or less (Painter and Bedoya 1991:15, 24–25).

Settlement of the Chapare began in the 1930s, and the population increased gradually into the mid-1970s. In 1967, 24,361 people resided in 54 settlements in the Chapare (Flores and Blanes 1984:82). Settlement accelerated rapidly in the mid-1970s, coinciding with an economic crisis in eastern Bolivia triggered by a precipitous drop in international cotton prices. Many farmers in Santa Cruz, responding to financial incentives provided by the Bolivian government and international donor agencies, had gone heavily into debt to expand cotton production, and the decline in cotton prices left them in dire financial circumstances. By some accounts (e.g., Bascopé 1982:53–56; Dunkerley 1984:222–223, 315), this crisis precipitated large-scale investment in coca leaf processing by parts of the Santa Cruz agricultural elite, and stimulated demand for coca leaf. By the end of 1981, there were 247 settlements in the Chapare, and the settler population was approximately 83,525 (Flores and Blanes 1984:88–89). This represented a 243 percent increase in the settler population and a 357 percent increase in the number of *colonias* in the Chapare (Jones 1990). By 1987, this population had more than doubled again, to between 196,000 and 234,000 (Durana et al. 1987).

Figure 2, below, draws on information collected from coca farmers by the Bolivian government's *Dirección de Reconversión Agrícola* (DIRECO), to confirm this general migratory pattern.[6] As may be seen, Chapare farmers are predominantly from within Cochabamba department. In fact, of 10,703 farmers interviewed by DIRECO, 6,867 (65.2 percent) were from Cochabamba. Potosí ranks second as the

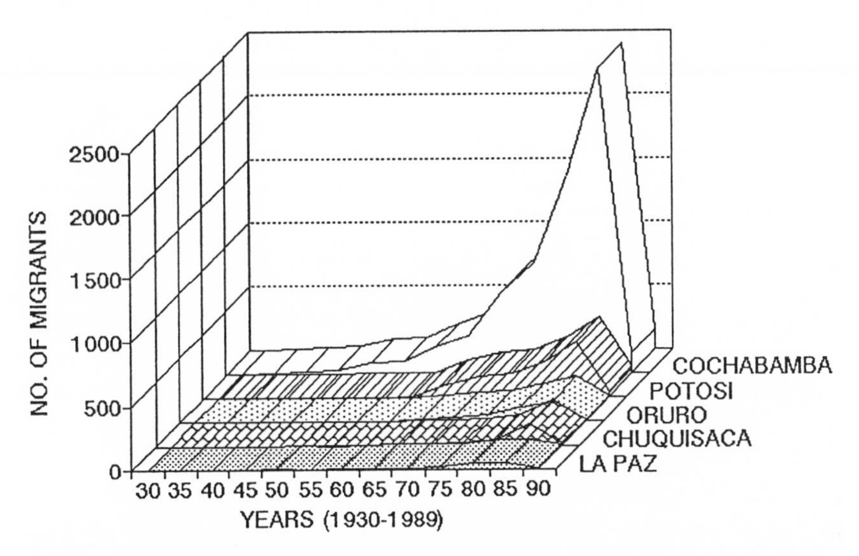

Fig. 2. Migration to the Chapare, by department of origin and year of arrival. (Data from DIRECO database.)

department of origin, accounting for 1,292 (12.3 percent) of the farmers interviewed by DIRECO.[7] The predominance of Cochabamba as place of origin contrasts sharply with the origins of farmers in the Yapacaní and San Julián settlement areas of northern Santa Cruz department. In Yapacaní, which borders the Chapare on the east (but which settlers reached by going first to the city of Santa Cruz and then moving northward) the departments of Cochabamba and Potosí are of nearly equal importance, with each accounting for approximately one-third of the settlers. The department of Chuquisaca is the next most important source of migrants to Yapacaní, accounting for approximately 19 percent. In San Julián, Potosí is the largest source of settlers, accounting for 44 percent, with Chuquisaca and Cochabamba ranking second and third, accounting for 20 and 15 percent of the settlers, respectively (Painter and Bedoya 1991:23).

The contrasting importance of departments of origin of settlers in different areas is suggestive of how the dynamic resulting in the settlement of the Chapare differed from other areas. Table 1 shows the province of origin of the Chapare farmers from Cochabamba department interviewed by DIRECO. As may be seen, the largest number come from Chapare prov-

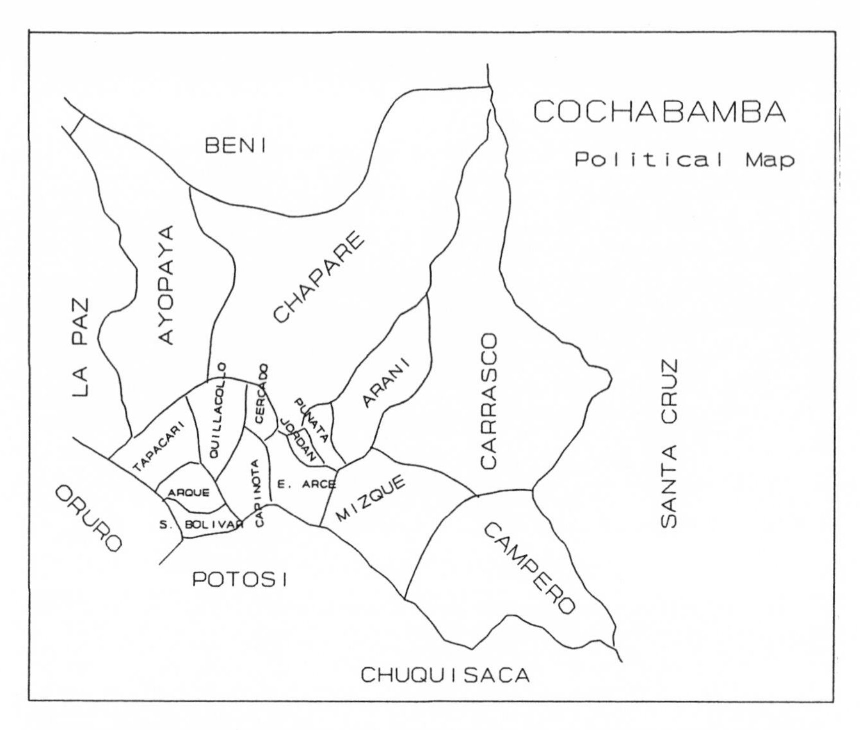

Fig. 3. Cochabamba

ince itself, and 70 percent come from provinces bordering Chapare province (see fig. 3). This suggests that proximity to the Chapare has been particularly important in farmers' decisions to settle there, especially when compared with the other settlement areas mentioned.

In addition to simple proximity, however, historical factors help explain the population movements in recent decades. The Chapare that has been the object of land settlement is an area composed of the tropical portions of Chapare and Carrasco provinces, which is drained by the Chapare and Chimoré-Ichilo rivers. While only these two provinces contain tropical areas as well as upland areas, much of the population of the neighboring upland provinces maintained access to land in different ecological zones, including tropical land, along the slopes leading down into the Chapare. This was the basis of Intendent Viedma's plan, in the 1780s, to build a road into the Chapare to promote commercial coca leaf production there on a scale that would compete with production from the *yungas* area north of La Paz (Larson 1988:246–258).

In fact, people have cultivated coca leaf and other tropical products on the lower slopes and in the Chapare for several centuries. The area never supplied as much coca leaf to the mines and other traditional consumption markets as did the *yungas*, but this production was important for family consumption, trade, and sale. In interviews, many farmers tell of large numbers of people leaving their homes beginning in the late 1970s, and particularly during the 1980s. In some extreme cases, one finds isolated communities that have been totally abandoned, with empty houses falling into disrepair. The drought of the 1980s and a lack of water more generally are often cited as the specific reason prompting the move (e.g., Jones 1990:18–19). The drought and economic traumas of the 1980s were for many the finishing blows that occurred within a context of long-term economic decline and land degradation, forcing many to abandon land altogether and many others to become significantly more dependent on off-farm income sources.

A high percentage of settlers come from the upland areas immediately bordering the Chapare. Farmer interviews and direct observation indicate a pattern of upland fields and residences being abandoned or falling into disuse. This suggests that many settlers abandoned the up-

TABLE 1. Provinces of Origin for Farmers in the Chapare from Cochabamba Department

Province	No. of Farmers	%
Chapare	1,981	28.8
Quillacollo	1,031	15.0
Arani	484	7.0
Ayopaya	483	7.0
Capinota	406	5.9
Punata	355	5.2
Arque	290	4.2
Tapacari	274	4.0
Cercado	249	3.6
Carrasco	214	3.1
Esteban Arze	194	2.8
Mizque	131	1.9
Campero	53	0.8
Jordan	51	0.7
Bolívar	2	0.0
No response	669	9.7
Total	6,867	99.7

Source: Data from Direco database.

land portion of what had been vertically organized production systems along the slopes leading from the uplands to the tropical Chapare. Table 2, for example, shows the communities of origin of farmers from Chapare and Carrasco provinces, the two provinces containing the tropical area commonly referred to as the Chapare. As may be seen, 91 percent are from upland localities. These upland areas have long-standing vertical production linkages of different types extending downward toward the tropical Chapare. It is probable that many of the other upland provinces bordering the Chapare have similar histories of agricultural production. The family labor required to sustain agricultural production in these areas was too great in the face of declining productivity and profitability. Many families apparently hoped to improve the return to their labor by concentrating their efforts in the tropical lowlands.

TABLE 2. Localities of Origin—Farmers from Chapare and Carrasco Provinces, Cochabamba Department

Locality	Province	No. of Farmers	%
Upland			
Sacaba	Chapare	635	41.4
Colomí	Chapare	552	36.0
Totora	Carrasco	100	6.5
Ucuchi	Chapare	54	3.5
Palca	Chapare	18	1.2
Pojo	Carrasco	13	0.8
Pocona	Carrasco	10	0.7
Corani	Chapare	5	0.3
Candelaria	Chapare	4	0.3
Chillichi	Carrasco	4	0.3
Quehuiñapampa	Chapare	3	0.2
Subtotal		1,398	91.3
Lowland			
Todos Santos	Chapare	44	2.9
Villa Tunari	Chapare	14	0.9
Espíritu Santo	Chapare	11	0.7
Chipiriri	Chapare	9	0.6
Paractito	Chapare	9	0.6
Chimore	Chapare	3	0.2
Ibirgarsama	Carrasco	1	0.1
Other	—	43	2.8
Subtotal		134	8.7
Total		1,532	100.0

Source: Data from DIRECO database.

Chapare farms are small compared to those in lowland settlement areas elsewhere in Bolivia, with the mean size between 9 and 12 hectares. Farm sizes in the San Julián area averaged 40 hectares, with farmers who formalized their settlement through the INC receiving 50 hectare lots (Painter et al. 1984). In Yapacaní, the mean farm size is approximately 29 hectares. The mean area of settlers' landholdings in the Chapare is about 10 hectares, with between 1.25 and 4 hectares under cultivation. Coca leaf accounts for between 0.5 and 1.6 hectares of the area under cultivation.

Even more important than the mean figures, however, is the unequal land distribution. Table 3 shows the distribution of land among the farmers in the DIRECO database. It further divides the farmer population according to residence in one of the Chapare's seven microregions, which are relatively homogeneous with respect to agricultural production conditions.[8] While the mean farm size is 10 hectares, 25 percent of the farmers who provided the information appearing in the table own less than five hectares. Table 4 looks at the same information another way, showing the distribution of land among different cohorts of farm-

TABLE 3. Frequency Distribution of Chapare Farms by Lot Size

Plot size (hectares)	Microregions						
	Reg. 1	Reg. 2	Reg. 3	Reg. 4	Reg. 5	Reg. 6	Reg. 7
0	23	37	178	79	14	7	2
0<n≤5	208	65	847	660	128	21	2
5<n≤10	237	211	1,597	1,385	112	93	14
10<n≤15	71	178	291	179	19	57	24
15<n≤20	30	688	513	93	4	294	49
20<n≤25	7	27	27	7	2	20	1
25<n≤30	1	2	7	3	1	0	43
30<n≤35	1	6	0	0	0	0	20
35<n≤40	1	3	6	2	0	1	9
40<n≤45	0	0	1	1	3	0	2
45<n≤50	0	8	0	0	0		19
50<n≤55	0	0	0	1	0		1
55<n≤60	0	4	1	0	0		9
60<n≤65	0	0	0		0		0
65<n≤70	0	0	1		1		
70<n≤75	0	1	0		0		
75<n≤80	1	0	2				
80<n≤85	0	1	0				
No. of farms	580	1,231	3,471	2,410	284	493	195
Max. farm size	78	82	80	54	70	36	60

Source: Data from DIRECO database.

ers, ranging from the 10 percent who own the least land to the 10 percent who own the most. The 20 percent of the settlers with the largest farms own about five and a half times the amount of land as the 20 percent of settlers with the smallest farms own.

One result of the large number of small landholdings, coupled with the variation in agricultural potential of Chapare land, is that the upland pattern of families relying heavily on off-farm income is reproduced in the Chapare. Various accounts of how families organize migration between upland and lowland areas (e.g., Carafa et al. 1987; CEFOIN 1990; Jones 1990) report that it is common for husbands and older sons to work family land in the Chapare while wives and daughters remain in upland areas, caring for small children, farming if the family maintains land in the upland area, or attending to some commercial activity. Various data suggest that off-farm income, most of which is generated by women, is very important for Chapare settlers. While only 40 percent of the farmers interviewed by CERES had land in their areas of origin, 66 percent owned a house or other property outside the Chapare. Also, the DIRECO information on settler household composition shows a singularly low number of female offspring. The total number of male offspring reported by the 10,703 farmers interviewed was 25,388, which works out to a mean of 2.37 male children per household. At the same time, farmers reported that only 3,213 female offspring resided with them, for a mean of 0.3 female children per settler household. Thus, more than eight times as many sons as daughters live in the Chapare.

These figures contrast sharply with others from the Yapacaní settlement area, which borders the Chapare to the east, but to which farmers arrived as part of the settlement of northern Santa Cruz, rather than through the Chapare. DIRECO interviewed 1,143 farmers in this area, who reported totals of 2,094 sons and 1,894 daughters residing with them. This works out to means of 1.83 sons and 1.62 daughters, respectively, per household, or only 1.13 times as many sons present as daughters.

The low number of female children per household appears to confirm the importance of women working outside of the Chapare to settler family incomes,[9] an importance undoubtedly due in part to the proximity of the Chapare to the upland areas from which most settlers come. Families can divide their efforts and conduct economic activities in upland areas as well as in the Chapare, but this decision is also an outcome

TABLE 4. Concentration of Landholdings in the Chapare

	Microregions													
	Reg. 1		Reg. 2		Reg. 3		Reg. 4		Reg. 5		Reg. 6		Reg. 7	
Strata	Area	%	Area	%	Area	%	Area	%	Area	%	Area	%	Area	%
Bottom 10%	41.5	0.90	352.0	1.80	289.3	0.88	341.5	1.68	9.0	0.47	202.7	2.55	123.50	2.35
9th 10%	159.1	3.46	1,092.1	5.59	1,275.1	3.87	1,065.2	5.26	57.0	2.95	491.4	6.18	245.00	4.66
8th 10%	249.0	5.41	1,361.2	6.97	1,741.0	5.29	1,205.0	5.94	96.1	4.97	565.4	7.11	338.00	6.43
7th 10%	290.0	6.30	1,780.0	9.11	2,357.6	7.16	1,616.6	7.98	120.8	6.24	755.0	9.49	384.00	7.31
6th 10%	372.3	8.09	2,077.3	10.64	3,177.5	9.65	2,284.4	11.27	145.0	7.49	938.4	11.80	429.00	8.16
5th 10%	486.3	10.57	2,300.5	11.78	3,470.0	10.54	2,410.0	11.89	178.0	9.20	1,000.0	12.57	550.70	10.48
4th 10%	574.0	12.48	2,460.0	12.60	3,470.0	10.54	2,410.0	11.89	215.4	11.13	1,000.0	12.57	599.00	11.40
3rd 10%	580.0	12.61	2,460.0	12.60	3,734.0	11.34	2,410.0	11.89	256.7	13.27	980.0	12.32	681.20	12.96
2nd 10%	686.8	14.93	2,460.0	12.60	5,999.2	18.22	2,473.6	12.20	290.5	15.01	980.0	12.32	888.00	16.90
Top 10%	1,161.1	25.24	3,185.4	16.31	7,417.9	22.53	4,053.8	20.00	574.4	29.69	1,039.7	13.07	1,017.30	19.36

Source: Data from DIRECO database.
Note: Area in hectares.

of the combination of small, fragmented landholdings in the Chapare and the extremely restricted economic opportunities in agriculture. Despite the importance of income from coca leaf, these problems oblige Chapare families to rely heavily on off-farm income, just as they do in impoverished upland areas.

Wage Labor Migration from Upland Areas

This discussion has focused on migration and settlement of the Chapare by farmer families from upland valleys. The rural upland population is not homogeneous, and has responded to the crisis of underdevelopment and land degradation in diverse ways. Primarily because of the importance of the area in producing coca leaf for cocaine, the Chapare provides wage labor for many people. Large numbers of poor rural people migrate to the Chapare for varying lengths of time to work in the care of coca plants and the harvesting of coca leaf, and many more work carrying coca leaf from isolated farms to bulking and processing centers and in the initial stages of extracting the cocaine alkaloid from coca leaf.

Farmers tend to be from the area immediately surrounding the Chapare, but the Chapare provides wage labor to people living in a much wider area of upland Bolivia. While only about 2 percent of Chapare farmers are from the provinces of Campero and Mizque, for example, the Chapare is the most frequent migratory destination for rural families in these two provinces when they seek off-farm income. As part of its research on the rural economy of Campero and Mizque, the *Centro de Formación Integral* (CEFOIN), of the University of San Simón, Cochabamba interviewed 46 elected leaders (*dirigentes*) of communities in the two provinces, and asked them about the most important migratory destinations of people from their communities.[10] Among them, the leaders listed 89 migratory destinations, of which the three most important were the Chapare (mentioned 38 times), Santa Cruz (mentioned 23 times), and the city of Cochabamba (mentioned nine times). Similarly, CEFOIN asked 60 returned migrants about their destinations and received 73 responses, of which 39 included the Chapare, 25 included Santa Cruz, and 9 included the city of Cochabamba (CEFOIN 1990:165–166; Pérez 1991a:28–29).

Pérez (1991a) notes that the general pattern is for wage laborers to migrate several times a year for relatively short duration, but the length of reported migration varies very widely, and almost all of the remaining migratory destinations mentioned that are ranked lower than the three most important are upland towns and cities within a short distance of migrants' homes. Since most migrants report several destinations where they go regularly to seek work, we see that migration as we conventionally think of it represents the extreme cases in a range of practices through which families earn off-farm income. This suggests that off-farm income is even more important to the survival of rural families than we might be led to believe by looking at migration data, and that, while migration data are dramatic, they may actually underrepresent the crisis facing upland agriculture.

TABLE 5. Frequency Distribution of Chapare Farms by Cultivated Area

Plot size (hectares)	Microregions						
	Reg. 1	Reg. 2	Reg. 3	Reg. 4	Reg. 5	Reg. 6	Reg. 7
0	4	6	23	8	2	2	1
0<N≤3	300	610	2,602	1,749	238	343	75
3<N≤6	183	453	677	565	36	112	63
6<N≤9	58	120	111	67	7	28	23
9<N≤12	25	26	36	15	0	1	12
12<N≤15	5	7	9	4	0	2	8
15<N≤18	2	2	8	1	0	3	4
18<N≤21	2	1	3	1	0	1	3
21<N≤24	1	1	0	0	1	1	2
24<N≤27	0	1	0		0	0	2
27<N≤30		0	1				1
30<N≤33		0	0				0
33<N≤36		0					0
36<N≤39		0					0
39<N≤42		1					0
42<N≤45		0					0
45<N≤48		0					0
48<N≤51		1					1
51<N≤54		1					0
54<N≤57		0					
57<N≤60		0					
60<N≤63		0					
63<N≤66		0					
66<N≤69		0					
69<N≤72		0					
72<N≤75		1					
75<N≤78		0					
No. of farms	580	1,231	3,470	2,410	284	493	195

Source: Data from DIRECO database.

TABLE 6. Frequency Distribution of Chapare Farms by Area in Coca Cultivation

Plot Size	Microregions						
(hectares)	Reg. 1	Reg. 2	Reg. 3	Reg. 4	Reg. 5	Reg. 6	Reg. 7
0	48	59	71	46	9	5	5
0<N≤1	277	398	1,825	1,145	189	210	129
1<N≤2	184	414	1,118	838	65	190	35
2<N≤3	51	227	363	278	13	63	14
3<N≤4	13	74	67	68	6	15	10
4<N≤5	4	38	17	17	1	9	1
5<N≤6	3	16	4	12	0	0	1
6<N≤7	0	3	2	2	1	1	0
7<N≤8		1	0	2	0	0	
8<N≤9		0	3	1			
9<N≤10		1	0	1			
10<N≤11		0	0	0			
No. of farms	580	1,231	3,470	2,410	284	493	195

Source: Data from DIRECO database.

TABLE 7. Total Areas in Hectares Dedicated to Crop Production by Chapare Farmers Included in DIRECO Database

	Microregions						
Crop	Reg. 1	Reg. 2	Reg. 3	Reg. 4	Reg. 5	Reg. 6	Reg. 7
Coca	605.87	1,937.44	3,758.44	2,936.97	249.68	632.61	176.62
Bananas	454.53	399.56	885.05	257.44	63.49	121.43	174.46
Cassava	107.11	308.47	609.06	527.91	36.61	97.90	55.95
Corn	52.42	202.76	171.68	75.43	11.02	34.88	32.08
Rice	140.90	724.73	1,121.81	1,063.72	37.22	159.93	345.52
Citrus	597.37	755.40	1,026.15	751.99	60.60	224.16	42.08
Taro	3.84	3.28	18.42	15.10	4.10	1.38	4.00
Avocado	59.96	6.63	52.24	21.53	2.49	1.78	2.13
Other 1	45.10	128.92	254.69	74.12	24.14	38.87	240.76
Other 2	6.60	21.85	26.26	12.78	6.01	4.98	2.94

Source: Data from DIRECO database.

The Organization of Lowland Agricultural Production

General Tendencies

As discussed above, the mean farm size in the Chapare is about 10 hectares, with the area under cultivation ranging between 1.25 and 4 hectares and the area dedicated to coca ranging between 0.5 and 1.6 hectares. The modal farm size is something less than 10 hectares, and the variation around the mean varies considerably from one microregion to another. Tables 5 and 6 summarize the distribution of Chapare farms according to the area of land under cultivation and the area of land dedicated to coca cultivation. As may be seen, the mode for the amount of land under cultivation is slightly less than two hectares, while the modal amount of land dedicated to coca leaf cultivation is slightly less than one hectare. Table 7 summarizes the mean area per farmer in each microregion dedicated to the major crops cultivated.[11]

In general, Chapare farmers are smallholders with a diversified cropping system and few animals. Coca is the single most important crop, as one would expect, since it is the crop for which there has been the most reliable market over the last decade, but coca is by no means the only crop cultivated by Chapare farmers. Chapare production systems include significant quantities of annual and perennial crops. In fact, in almost all cases, the amount of land dedicated to annual crops and other perennial crops exceeds the amount of land under coca cultivation. Coca amounts to approximately 40 percent of the area under cultivation in the Chapare.[12]

Crop production in the Chapare depends on unwaged family labor. Of the 194 farmers interviewed by CERES, 94 percent indicated that farming is their major occupation, and only 13 percent stated that they ever work off-farm. The 194 farmers interviewed by CERES were accompanied by 343 family members, 70 percent of whom stated that their major occupation was unremunerated work on the family farm. While 58 percent of the farmers interviewed by CERES stated that they do hire labor from time to time, 87 percent said they regularly use unremuner-

TABLE 8. Type of Labor Force Utilized According to Land Area under Cultivation

Number of Cultivated Hectares	Only Family Labor Force		Family Labor Plus Hired Labor		Total	
	No. of Farms	%	No. of Farms	%	No. of Farms	%
0.1 to 2.99	28	67.0	14	33.0	42	100.0
3 to 4.99	16	36.0	29	64.0	45	100.0
5 to 6.99	4	25.0	12	75.0	16	100.0
7+	1	7.0	13	93.0	14	100.0
Total	49	42.0	68	58.0	117	100.0
					7	No Answer
					(*N* = 124)	

Source: Data from CERES interviews.
Note: 124 farmers were interviewed; 7 did not answer the question.

ated family labor, and 72 percent said they regularly practice *ayni* labor exchanges on their farms. Farmers regularly stated that the availability of unpaid family labor and the costs of hired labor are major constraints on their ability to produce.

As may be seen in table 8, hired labor is an essential part of farming in the Chapare, but two-thirds of the farmers interviewed with less than 2.99 hectares under cultivation—the majority of Chapare farmers—rely exclusively on family labor to perform agricultural tasks. At this level of production, hired labor is clearly a complement to the labor provided by family members. As we would expect, the importance of hired labor increases significantly among larger farmers, although family labor continues to be important.

Hired labor is used more in coca production than it is for crops in general, reflecting coca leaf's importance as farmers' most reliable source of cash. Table 9 shows that the smaller coca producers interviewed by CERES, those with one hectare or less, hire workers to supplement family labor resources more frequently than do farmers with larger areas of coca. Coca leaf is the most difficult crop to accommodate within a production system based on family labor, because its labor requirements are very high in comparison to other crops. The most important labor requirements are associated with the heavy amount of weeding required to protect young coca plants until they are big enough to shade the ground. One hectare of coca requires about 120 person/days in weeding, compared to 36 person/days for coffee, 32 for citrus,

TABLE 9. Type of Labor Force Utilized According to Land Area in Coca Leaf Cultivation

Number of Coca Hectares	Only Family Labor Force		Family Labor Plus Hired Labor		Total	
	No. of Farms	%	No. of Farms	%	No. of Farms	%
0.1 to 0.5	17	47.0	19	53.0	36	100.0
0.51 to 1.0	10	38.0	16	62.0	26	100.0
1.01 to 2.0	19	56.0	15	44.0	34	100.0
2.01+	1	7.0	16	94.0	17	100.0
Total	47	42.0	66	58.0	113	100.0
					11	No Answer
					(N = 124)	

Source: Data from CERES interviews.
Note: 124 farmers were interviewed; 11 did not answer the question.

and 24 for bananas (Henkel 1971:190, 210). In addition, when coca leaf is being grown for the purpose of extracting the cocaine alkaloid, harvesting and transport make particular demands on labor resources. If coca leaf is not harvested and processed within an optimal period that lasts only a few days, the alkaloid content of the leaf drops substantially (Henkel, personal communication).

The patterns of family and hired labor use in the cultivation of coca leaf and other crops suggests that the economic situation of Chapare farmers is precarious. Both from farmer comments in interviews and from the greater area dedicated to cultivating coca leaf than to other cultivars, we see that coca is the most marketable crop. It is also clear that coca leaf is relatively more important for smaller farmers than for larger ones, because of the relatively larger investment that small farmers make to establish a coca plantation.

At the same time, we see from the very weak relationship between the amount of land that a family owns and the area under coca cultivation, that farmers are not maximizing coca production. Using 10,581 cases in the DIRECO database that contained information on farm size and area under coca cultivation, we calculated two correlation coefficients to measure the strength of the relationship between the two variables. The first, using Pearson's r, yielded a correlation coefficient of .12842, showing a weak relationship between the size of each farmer's holding and the area under coca leaf cultivation. The second, using Spearman's r_s, sought simply to see if there is a relationship between the ranks of the farms according to size of landholding and their ranks according to the amount of land under coca cultivation. This yielded a coefficient of .16252, also indicating a weak positive relationship between the ranks. The relationship between the amount of coca cultivated and the amount of family labor available is weak as well. The Pearson's r coefficient of the relationship between the number of offspring a farmer has and the area under coca cultivation is .17092, while the Spearman's r_s coefficient of the relationship between the ranks of the farms according to number of offspring and area under coca cultivation is .21311.

Farmers are limiting the area they dedicate to coca leaf production according to considerations that are not particularly related to land and labor availability. Two explanations suggest themselves. One is that there is a system of quotas on production to exercise some control over price and ensure that coca revenues are broadly distributed among farmers. Such systems exist in many areas of the world. Peanut quotas enforced by the federal government in the southeastern United States and quotas on high value cash crops like tea, enforced by contract farming schemes, are examples that come to mind. One can imagine that the coca leaf producers' unions (*sindicatos*) might be interested in quotas to distribute the revenues from coca production among their members, while drug traffickers might be interested in dispensing production quotas as a means of exercising control over producers. In fact, it is unlikely that an explicit quota system—formal or informal—exists in the Chapare; certainly our research uncovered no evidence of one. However, approaching the question of what shapes coca leaf production levels in this light is suggestive of the kinds of extra-economic interests that may influence producers' decisions.

A more likely explanation, based on the information presented here, is that more coca is not being grown because it is not profitable in a formal sense. If coca leaf production were profitable, we would expect farmers to hire the labor necessary to expand it more rapidly. The Chapare would become almost exclusively a coca leaf producing area, and coca leaf production would be significantly constrained by land availability. Instead, farmers continue to rely heavily on the family obligations and labor exchanges described by Weil (1989) and others on the basis of fieldwork conducted in the early 1970s. In recent years, the price of coca has been very volatile, to a certain extent responding to cocaine and coca leaf repression efforts, and to a greater extent responding to pressures exerted at different levels of the narcotics industry that we simply do not understand (Jones 1990:13–15, 23–25). Farmers do not grow coca leaf because it is particularly attractive; they grow it because, even though the price is volatile and often low, it is the only crop for which they know they will have a buyer (Jones 1990; Rivera 1990).

The fact that farmers' most important crop, coca leaf, is not sufficiently profitable to provide incentives to hire labor and expand coca production is symptomatic of the kinds of pressures farmers are experiencing to use land extensively and destructively. Accompanying the lack of incentive to hire labor is a lack of incentive to invest in most capital inputs for the production of other crops. CERES found that only 5 percent of 124 farmers who provided information on this topic used improved seeds, and only 15 percent used fertilizers (Painter and Bedoya 1991:63). These figures reflect the kinds of trends in the Bolivian rural economy that caused coca leaf production to assume the prominent place it has. In 1976, despite primitive production technology, Chapare farmers produced 39 percent of Bolivia's plantains, 32 percent of its citrus, and 20 percent of its rice, at a time when only 6 percent of the area was occupied (OAS 1984:191). Today, farmers report that they cannot usually find a buyer for these products at any price, and their primary destination is consumption by farmers, their families, and any hired laborers. Many farmers give food in addition to or in lieu of cash wages to laborers, another symptom of the economic bind in which they are caught.[13]

At the same time, farmers are using large quantities of insecticides; 89 percent of those interviewed by CERES reported that they use insecticides. Farmers also report that the need for insecticides is increasing and that insecticides appear to be less effective than they used to be, an indication of broader land management problems. On the basis of the information from CERES's farmer interviews, the mean area per farmer of land in fallow is 6.39 hectares, approximately twice the 3.2 hectares that are the mean area under cultivation per farmer (Painter and Bedoya 1991:61–62). In addition, only 29 percent of the farmers interviewed by CERES indicated that they were managing fallow land by growing perennial crops on it; 71 percent of the farmers interviewed had agricultural systems that consisted of annual crops being rotated within an extensive fallow area, with no management of succession of annual crops through the use of tree crops. Furthermore, the percentage of farmers who use permanent crops to manage succession in fallow land declines as plot

TABLE 10. Principal Cropping System According to Plot Size

Plot Size (hectares)	Agricultural System with Fallow (Annual Crops + Fallow Area)		Agricultural System without Fallow (Permanent Crops)		Total	
	No. of Farms	%	No. of Farms	%	No. of Farms	%
0.1 to 5	20	53.0	18	47.0	38	100.0
5.1 to 10	46	68.0	22	32.0	68	100.0
10.1 to 20	32	94.0	2	6.0	34	100.0
20.1+	7	100.0	—	—	7	100.0
Total	105	71.0	42	29.0	147	100.0
					47	No Data
					194	

Source: Data from CERES interviews.
Note: 194 farmers were interviewed; 47 did not answer the question.

size increases (table 10), suggesting that larger farmers are more extensive in their use of land than small farmers.[14]

The lack of economic opportunity associated with Chapare farming discourages the use of modern inputs and leads farmers to rely heavily on family labor. In most cases, family labor is supplemented by hired labor only for coca leaf production, because of the role this crop plays in providing a minimum level of cash income for farm families. Even here, however, the level of profitability is insufficient to encourage producers to maximize coca leaf production. This suggests that coca leaf would not be a profitable crop if it were not for the extensive use of unwaged family labor. It is not grown because it is profitable, but because it is the only crop that farmers have reasonable assurance of being able to sell. The cash received from coca leaf is an increasingly tattered safety net, whose importance is greatest for the smallest farmers.

Since they are unable to earn sufficient revenues from farming to invest in inputs or labor, farmers seek to meet minimal production requirements through an extensive use of land. Lack of fertilizer use means that large areas of fallow must be maintained in order to maintain what has become an annual crop regimen directed almost exclusively to home consumption. Poverty and labor scarcity combine to make difficult the management of fallow areas through the use of perennial crops, for the purpose of maintaining soil fertility and increasing and diversifying farm production. Denevan and Padoch (1988:23–26, *passim.*) point out that some form of fallow management is a critical feature of the agricultural systems of native people and peasants in the tropical lowlands of South America, systems that have proven to be relatively stable for extended periods of time. The exact nature of this fallow management depends on a number of factors including land availability, soil conditions, and the way production is linked to regional, national, and international markets. More needs to be known about ways in which systems of fallow management might improve land use practices in the Chapare, but land management practices cannot improve without major increases in the incomes settlers receive from farming.[15]

Implications and Conclusions

The upland valleys of central Bolivia have long suffered from economic stagnation and underdevelopment, associated with the country's dependence on supplying export markets with a small number of commodities. In recent decades, for example, the economy has rested on minerals, particularly tin, and cocaine. These long-term problems were exacerbated by a series of crises that occurred during the 1970s and 1980s. The net effect on the rural population has been the establishment and acceleration of mutually reinforcing cycles of impoverishment and land degradation.

The major response to this has been migration, characterized by a wide variety of destinations and migratory patterns of diverse organization and duration. In the absence of investment in upland agriculture to improve market conditions and the productivity of family labor, migration transforms labor scarcity from a seasonal problem of peasant farmers to a chronic, year-round condition that obliges people to become increasingly destructive in their use of land. Because migrants are predominantly men, the principal agents and victims of land degradation are women and small children, and patterns of gender-based discrimination and oppression further complicate efforts to assist them in breaking the cycle of impoverishment and land degradation. This cycle actually increases the hardships to which women are subjected. As women are responsible for upland production that yields less and less in quantity and value, their economic contribution declines, and with it, their claim on political participation.

Chapare settlement was a specific outcome of the crisis of upland agriculture, and specifically of the collapse of vertical production systems linking upland and lowland areas along the slopes bounding the tropical lowlands. However, the same conditions that limited the production of food crops in upland areas choked the production of lowland food crops as well. The growing reliance on coca leaf production reflected the lack of opportunity in food crop production, but coca leaf has not improved the overall situation of Chapare farmers, because they provide a commodity whose price is unaffected by the labor embodied in it. As a result, Chapare farmers are encouraged to use land extensively and destructively, mining it as a substitute for capital to subsidize the costs of labor they are unable to pay.

NOTES

The research described in this chapter has been supported by the Cooperative Agreement on Systems Approaches to Regional Income and Sustainable Resource Assistance No. DHR 5452-A-00–9083–00 at Clark University, the Institute for Development Anthropology, and Virginia Polytechnic Institute, funded by the U.S. Agency for International Development, Bureau for Science and Technology, Office of Rural and Institutional Development, Division of Rural and Regional Development. The views and interpretations are those of the author and should not be attributed to the Agency for International Development or any individual acting on its behalf. While solely responsible for any errors of fact or interpretation, the author thanks Eduardo Bedoya, Deborah Caro, James Jones, and Carlos Pérez for their participation in the collection and analysis of the material discussed and for many helpful comments on portions of this chapter.

1. For example, following short-term declines in food production resulting from the disruptions of the revolution and land expropriation under the agrarian reform, production of most basic foodstuffs rose steadily between 1955 and 1960. However, the value of agricultural production, calculated as part of Bolivia's Gross Domestic Product, declined from $113.1 million in 1952 to $112.2 million in 1958. Farmers worked harder and produced more, but received less. Furthermore, the economic crisis of the 1950s, and the stabilization program it triggered, led to heavy Bolivian dependence on foreign assistance. In 1959, for example, $10 million of the $32 million received by the Bolivian government in revenue collections was from counterpart funds. Only 12 percent of U.S. financial assistance to Bolivia went to agriculture, and only a small portion of this was targeted to benefit smallholding farmers in upland areas. In fact, the United States spent more between 1954 and 1957 ($56.6 million) to finance imports of surplus agricultural commodities to Bolivia than it did between 1952 and 1972 ($40 million) to support agricultural development (Frederick 1977:40, 46, 261–262).

2. Not all of the lowland settlement areas are agricultural. For example, many of the ex-miners who went to Pando are involved in gold mining cooperatives.

3. These are the same areas, discussed above, that are being converted to agricultural use as a result of land pressure.

4. Recent research (e.g., CEFOIN 1990:vii, 8, 66) has provided quantitative documentation of current herd sizes and composition in different areas of Campero and Mizque. However, there is no reliable quantitative information on changes in the numbers and kinds of animals during the 1980s.

5. This material will henceforth be referred to as the CERES interviews. The overall findings of the study are summarized in Rivera (1990). The figures cited here result from a retabulation of the interview material conducted by Eduardo Bedoya. CERES interviewed a nonrandom sample of 194 farmers who were selected as representing characteristics of the farmer population that the research team felt to be of interest, including location of land in one of seven ecological microregions, size of landholding, length of residence, and family size. The insights offered by the CERES dataset and its limitations are discussed in Painter and Bedoya (1991).

6. This material, henceforth referred to as the DIRECO database, consists of information collected from 10,703 Chapare farmers who discussed, with

DIRECO officials, their voluntary reduction of the amount of coca they were producing between November 1985 and the end of December 1989. The material remained in unanalyzed questionnaires until it was computerized and analyzed as part of the research being discussed here. The dataset represents a nonrandom sample of between 20 and 40 percent of the Chapare farmer population. The quality of the information and limitations on drawing conclusions from it are discussed in Painter and Bedoya (1991).

7. Close readers may notice that 6,867 is 65.2 percent of 10,531, rather than the 10,703 cases that comprise the data base. Because the material was collected over an extended period of time, with no analysis, and the collectors had unclear or contradictory notions about the purpose of the data collection, the format of the questionnaires varied and there were many inconsistencies in how responses were recorded, resulting in anomalous or missing information in the database.

8. The Chapare has been divided into seven microregions by the *Instituto Boliviano de Tecnología Agropecuaria* (IBTA), of the Bolivian government, largely on the basis of soil characteristics. The microregions vary considerably with respect to their potential for supporting agriculture, with Microregions I and III generally regarded as having the greatest agricultural potential.

9. An unfortunate failing of the DIRECO data is that they were gathered under the assumption that all farmers were male, and do not disaggregate farmers by sex. Also, the DIRECO data provide no information on farmers' spouses. These shortcomings necessitate the rather roundabout inference presented here. Jones (1990) found evidence of a significant minority of women farmers in the Chapare, although he was not able to make any quantitative assessment.

10. This study was conducted by CEFOIN in collaboration with the broader research effort being described here. The results are summarized in CEFOIN (1990). The survey information collected by CEFOIN was subsequently reanalyzed by Melissa Cable, on whose work the figures presented here are based. Pérez (1991a) has analyzed the rural economy of Campero and Mizque, drawing heavily on the CEFOIN report and Cable's reanalysis of the survey materials.

11. Many will undoubtedly wonder about the distinction between the "Other 1" and "Other 2" categories in table 7. It was clearly meaningful to the people who originally designed the DIRECO interviews. But, over the years, its significance was lost, and interviewers began to use one or the other arbitrarily to indicate crops not previously listed. After some discussion with colleagues on our research team, I have elected to maintain the labeling distinction.

12. DIRECO figures place coca production at approximately 44 percent of a farmer's cultivated area, on average, while the data collected by CERES yield average figures that range between 37 and 39 percent of the area under cultivation. The difference in the results of the two studies is not significant.

13. The wives of men who go to the Chapare as laborers report that, because of the declining production and productivity of upland farming systems, the food that men bring home from the Chapare occupies an increasingly central place in satisfying families' basic consumption needs (Gisbert et al. 1994). This is indicative both of the threat to the subsistence production regi-

men that we conventionally think of as subsidizing participation in the wage labor market, and of the growing inadequacy of the wages that workers are able to bring home.

14. Although the DIRECO database contains information on areas of land in fallow and forest, this discussion relies on the material contained in the CERES interviews. The DIRECO materials appear to reflect accurately the areas dedicated to individual crops. The DIRECO figures for fallow and forest are inconsistent with the information on individual crops, however, and appear unreliable. The intensive CERES interviews, on the other hand, do appear to be accurate. The discussion of the data gathered by CERES draws on conclusions reached by Eduardo Bedoya in the analysis he conducted as part of his contribution to Painter and Bedoya (1991).

15. This is not intended to imply that better fallow land management in the Chapare is the key to reversing land degradation. It was the focus here because it indicates the land management problems arising from the poverty of Chapare farmers. Problems of land degradation in the Chapare are more complex, beginning with the fact that only a small part of the region is suitable for prolonged farming, because of heavy rainfall and steep slopes.

REFERENCES

Bakx, K.
1987 Planning Agrarian Reform: Amazonian Settlement Projects, 1970–86. Development and Change 18(4):533–555.

Balán, J., and J. Dandler
1986 Marriage Process and Household Formation: The Impact of Migration on a Peasant Society. Report Prepared for the Population Council. La Paz: N.p. Manuscript.

Bascopé Aspiazu, R.
1982 La Veta Blanca: Coca y Cocaina en Bolivia. La Paz: N.p.

Blanes, J.
1984 De los Valles al Chapare. Cochabamba: CERES.

Carafa, Y., S. Arellano, and M. Uribe
1987 Tratamiento de la Temática de la Mujer en los Valles del Sur de Cochabamba. Report Prepared for USAID/Bolivia in Support of the Redesign of the Chapare Regional Development Project. La Paz: U.S. Agency for International Development.

CEFOIN (Centro de Formación Integral)
1990 Monografía del Distrito Sur. Report prepared by the Centro de Formación Integral, Universidad Mayor de San Simón, for the Programa de Desarrollo Alternativo Regional. Cochabamba: PDAR.

CIDRE (Centro de Investigación y Desarrollo Regional)
1987 Monografía de la Provincia de Mizque. Cochabamba: Centro de Investigación y Desarrollo Regional.

Collins, J. L.

1986 Smallholder Settlement of Tropical South America: The Social Causes of Ecological Destruction. Human Organization 45 (1):1–10.

1987 Labor Scarcity and Ecological Change. *In* Lands at Risk in the Third World: Local-Level Perspectives. P. D. Little and M. M Horowitz, eds. 17–37. Boulder, Colo.: Westview Press.

Crabtree, J., G. Duffy, and J. Pearce

1987 The Great Tin Crash: Bolivia and the World Tin Market. London: Latin America Bureau (Research and Action), Ltd.

Cuba, P.

1989 El Crédito Agrícola en el Distrito Sur: Una Consideración del Crédito para el Pequeño Productor Campesino. Cochabamba: CERES Manuscript.

Dandler, J.

1984 Campesinado y Reforma Agraria en Cochabamba (1952–1953): Dinámica de un Movimiento Campesino en Bolivia. *In* Bolivia: La Fuerza Histórica del Campesinado. F. Calderón and J. Dandler, eds. Pp. 201–239. Cochabamba: CERES and UNRISD.

Deere, C. D., and R. Wasserstrom

1980 Household Income and Off-Farm Employment among Smallholders in Latin America and the Caribbean. Paper Presented to the Seminario Internacional sobre la Producción Agropecuaria y Forestal en Zonas de Ladera in América Latina. Sponsored by CATIE. December 1–5. Turrialba, Costa Rica.

Denevan, W. M., and C. Padoch, eds.

1988 Agroforestería Tradicional en la Amazonía Peruana. Documento 11. Lima: CIPA.

Dunkerley, J.

1984 Rebellion in the Veins: Political Struggle in Bolivia, 1952–1982. London: Verso Edition.

Dunkerley, J., and R. Morales

1986 Crisis in Bolivia. New Left Review 155:86–106.

Durana, J., N. Anderson, and W. Brooner

1987 Population Estimate for the Chapare Region, Bolivia. Washington, D.C.: Development Strategies for Fragile Lands.

Flores, G., and J. Blanes

1984 ¿Donde va el Chapare? Cochabamba: CERES.

Frederick, R. G.

1977 United States Aid to Bolivia, 1953–1972. Ph.D. dissertation. Government and Politics Department. College Park: University of Maryland.

Gill, L.

1987 Peasants, Entrepreneurs, and Social Change: Frontier Development in Lowland Bolivia. Boulder, Colo.: Westview Press.

Gisbert, M. E., and M. Quitón

1991 Mujer y Migración en las Provincias de Campero y Mizque. IDA

Working Paper 92. Binghamton, N.Y.: Institute for Development Anthropology.

Gisbert, M. E., M. Painter, and M. Quitón
1994 Gender Issues Associated with Labor Migration and Dependence on Off-Farm Income in Rural Bolivia. Human Organization 53 (2):110–122.

Harris, O., and X. Albó
1986 Monteras y Guardatojos: Campesinos y Mineros en el Norte de Potosí. La Paz: CIPCA. (Originally Published by CIPCA in 1976.)

Healy, K.
1991 Structural Adjustment, Peasant Agriculture and Coca in Bolivia. Paper presented to the XVI International Congress of the Latin American Studies Association, April 4–6. Washington, D.C.

Heath, D. B.
1969 Land Reform and Social Revolution in Eastern Bolivia. *In* Land Reform and Social Revolution in Bolivia. H. C. Beuchler, C. Erasmus, and D. B. Heath. New York, N.Y.: Praeger Publishers.

Henkel, R.
1971 The Chapare of Bolivia: A Study of Tropical Agriculture in Transition. Ph.D. dissertation. Geography Department. Madison: University of Wisconsin.

Hess, D.
1980 Pioneering in San Julian: A Study of Adaptive Strategy Formation by Migrant Farmers in Eastern Bolivia. Ph.D dissertation. Department of Anthropology. Pittsburgh, Penn.: University of Pittsburgh.

Hjort, A.
1982 A Critique of "Ecological" Models of Pastoral Land Use. Nomadic Peoples 10:11–27.

Jones, J. C.
1990 Farmer Perspectives on the Economics and Sociology of Coca Production in the Chapare. IDA Working Paper 77. Binghamton, N.Y.: Institute for Development Anthropology.

Larson, B.
1988 Colonialism and Agrarian Transformation in Bolivia: Cochabamba 1550–1900. Princeton, N.J.: Princeton University Press.

OAS (Organization of American States)
1984 Integrated Regional Development Planning: Guidelines and Case Studies from the OAS Experience. Washington, D.C.: Department of Regional Development, Secretariat for Economic and Social Affairs, Organization of American States.

Painter, M.
1987 Unequal Exchange: The Dynamics of Settler Impoverishment and Environmental Destruction in Lowland Bolivia. *In* Lands at Risk in the Third World: Local-Level Perspectives. P. D. Little and M. M Horowitz, eds. 164–191. Boulder, Colo.: Westview Press.

1990 Development and Conservation of Natural Resources in Latin America. *In* Social Change and Applied Anthropology: Essays in Honor of David W. Brokensha. M. S. Chaiken and A. K. Fleuret, eds. 231–245. Boulder, Colo.: Westview Press.

Painter, M., and E. Bedoya Garland

1991 Socioeconomic Issues in Agricultural Settlement and Production in Bolivia's Chapare Region. IDA Working Paper 70. Binghamton, N.Y.: Institute for Development Anthropology.

Painter, M., and W. L. Partridge

1989 Lowland Settlement in San Julian, Bolivia—Project Success and Regional Underdevelopment. *In* The Human Ecology of Tropical Land Settlement in Latin America. D. A. Schumann and W. L. Partridge, eds. 340–377. Boulder, Colo.: Westview Press.

Pérez Crespo, C. A.

1987 San Julián: Balance y Desafíos. *In* Desarrollo Amazónico: Una Perspectiva Latinoamericana. Centro de Investigación y Promoción Amazónica and Instituto Andino de Estudios de Población, eds. Pp. 185–207. Lima: CIPA and INANDEP.

1991a Migration and the Breakdown of a Peasant Economy in Central Bolivia. Working Paper 82. Binghamton, N.Y.: Institute for Development Anthropology.

1991b Why do People Migrate? Internal Migration and the Pattern of Capital Accumulation in Bolivia. Working Paper 74. Binghamton, N.Y.: Institute for Development Anthropology.

Rivera, A.

1990 Diagnóstico Socio-económico de la Población del Chapare. Report Prepared for the Programa de Desarrollo Alternativo Regional. Cochabamba: PDAR.

Schmink, M., and C. H. Wood

1987 The "Political Ecology" of Amazonia. *In* Lands at Risk in the Third World: Local-Level Perspectives. P. D. Little and M. M Horowitz, eds. 38–57. Boulder, Colo.: Westview Press.

Stearman, A. M.

1976 The Highland Migrant in Lowland Bolivia: Regional Migration and the Department of Santa Cruz. Ph.D. dissertation. Anthropology Department. Gainesville: University of Florida.

1985 Camba and Kolla: Migration and Development in Santa Cruz, Bolivia. Gainesville: University Presses of Florida.

Urioste F. de C., M.

1989 Resistencia Campesina: Efectos de la Política Económica Neoliberal del Decreto Supremo 21060. La Paz: Talleres CEDLA.

USAID/MACA (U.S. Agency for International Development and Ministerio de Asuntos Campesinos y Agricultura)

1981 Encuesta Agropecuaria de la Región de los Valles del Sur de Bolivia. La Paz: U.S. Agency for International Development and Ministerio de Asuntos Campesinos y Agricultura.

Weil, C.
1989 Differential Economic Success among Spontaneous Agricultural Colonists in the Chapare. *In* The Human Ecology of Tropical Land Settlement in Latin America. D. A. Schumann and W. L. Partridge, eds. 264–297. Boulder, Colo.: Westview Press.

Whiteford, S.
1981 Workers from the North: Plantations, Bolivian Labor, and the City in Northwest Argentina. Austin: University of Texas Press.

Wolf, E.
1972 Ownership and Political Ecology. Anthropological Quarterly 45: 201–205.

Chapter 5

Environmental Destruction, Ethnic Discrimination, and International Aid in Bolivia

James C. Jones

This chapter relates environmental destruction in the Bolivian lowlands to socioeconomic and political forces emanating from the region and beyond. Abetting the forces, when not their essence, is ethnic discrimination. The chapter's historical narrative through 1979 focuses on Indian peasants in south-central Beni Department. They, like most peasants native to central and southern Beni, descend from mission Indians of Spanish colonial times, when Beni was the Jesuit Province of Moxos.[1] The Jesuits were the first Europeans to sustain contact with these peoples, whose post-Jesuit encounters with Bolivia's dominant Hispanic elite typify those of lowland Indians.

The chapter opens by sketching Beni's extreme physical environment; then, for a little-known part of South America, depicts historical periods, from the first Jesuit mission in 1667 to the eve of the 1952 revolution. This synopsis reveals time-worn elite economic behavior and attitudes toward Indians—both formidable barriers to environmental protection and native welfare. Even more revealing are events between the late 1940s and 1980, a dynamic period defined by three forces: expansive cattle ranching, which drove the regional economy of the period as wild rubber had driven it more than a half-century earlier; the 1953 Agrarian Reform; and the trade in peltry from the region's prolific fauna. These interlocking forces combined their furies to threaten both native society and the environment.

These forces vanished in the 1980s amid severe regional and national economic turmoil. New forces dominated the period 1980–1993, among them commercial logging and an Indian movement spawning new organizations and demands for native territories. The chapter details these and other period forces and official responses to them, while expanding the heretofore central-Beni focus to embrace all the *Oriente*, or eastern lowlands. It next questions official commitment to either environmental protection or indigenous welfare, and then ponders the cultural, policy, and donor-aid backdrop to environmental decline and ethnic discrimination. Finally, before closing, it offers counsel to international donors.

Especially relevant is a discussion of international aid, which today pours into Bolivia in apparent response to attractive, well-packaged policies and programs that suggest official concern for the environment and native welfare. Field observations and close analysis, however, belie this concern and give cause for alarm.

The chapter, whose scope is broad, provides a framework for considering environmental and indigenous welfare issues now on Bolivia's agenda. Many of the issues, if not the framework, are relevant elsewhere. The chapter seeks to teach at least two lessons: first, environmental destruction and racism closely link. Second, "development" has too often deepened both.

Physical Beni: Land of Extremes

Beni falls entirely within the Amazon Basin, as does 66 percent of the "Andean" Republic of Bolivia (Bolivia 1990:16). This poorly studied region is a sprawling web of forest, *pampa* (savannah), river, and lake. Extensive *pampa* dominates the central zone, where gallery forest flanks the rivers before flaring out exuberantly toward their western and southwestern headwaters to mantle the lower reaches of the eastern cordillera. The Chimán Forest, lying to the west and southwest of San Ignacio de Moxos, is part of this mantle. To the south of San Ignacio, beyond the cattle towns of San Lorenzo and San Francisco de Moxos, the *pampa* gives way to forest that grades gently upward through the Isiboro-Sécure National Park and into the Chapare zone of Cochabamba Department.

Beni's rivers are numerous and shallow, and because of the land's low gradient meander greatly as they course slowly northward across the plain. Channels shift dramatically, leaving a wide, plashy trail of oxbow lakes rich in aquatic fauna. On the plain, multitudinous lakes and lakelets accent the unrelenting flatness. The region's three river systems, the Beni, the Iténez, and the Mamoré, converge in the far north to flow into the Madeira through a narrow gap in the Brazilian Shield. Owing to this small egress, and to shallow beds and a low gradient, an enormous volume of water spills yearly onto the surrounding *pampas*. From January through March, parts of Beni appear from above as a giant lake relieved only by slips of gallery forest and small patches of hummock forest, known locally as *islas* (islands), scattered randomly across the savannahs.

The floods begin to recede in early April, leaving a myriad of sloughs and ponds, some teeming with fish. By mid-August, the peak of the winter dry season, drought conditions often exist: *pampa* grasses have browned and seared, and what were swales six months before are now parched flats of cracked, yellow earth.

The yearly cycle of flood and drought has long conditioned local life. At full flood, land animals move to scarce high ground, then gradually disperse as the waters recede, and finally gather again at water sources during drought. In Jesuit times, severe flooding forced the relocation of mission towns. Today, flooding makes refugees of residents in low-lying barrios of Trinidad, Beni's capital, and of peasants farming in alluvial levee soils along the River Mamoré.

Beni soils limit agriculture. *Pampa* soils are nutrient deficient and often underlain by a claypan. Gallery forest soils, of flood-borne alluvia, are better (Cochrane 1973:292–94; Denevan 1966:13; Muñoz Reyes 1977:86–89). Peasants practice slash-and-burn agriculture on forested ground, much of which is liberally pitted and thus unsuitable for farming during the rainy season when the pits fill with water. No one cultivates the *pampas* now, though land forms suggest someone did long before Europeans arrived.

Sures, or southern fronts, further enhance the land's asperity. They last from a day to a week and lash the region with cold winds, sometimes rain, during the winter dry season. These Patagonian winds can reach gale force—surface winds of 60 miles per hour have been reported (Denevan 1966:11)—and drive temperatures to four or five degrees centigrade.

Jesuit Moxos, 1667–1767

Between 1667 and 1767, Beni was the Jesuit Province of Moxos, where the Society of Jesus erected a network of some 25 *reducciones*, or mission towns, gathering into them village Indians of varied ethnic groups to instruct them in religion, politics, the arts, and European technologies and economic forms. Evidence points to small premission native settlements with populations that dispersed often to hunt, fish, and gather. Father Marbán says there were five or six villages every five or six leagues (20–25 km) (Denevan 1966:58). Father Castillo says the Moxos proper lived in some 70 villages, with an average population per village of 60 to 80 persons (Métraux 1942:54). The geographer William Denevan concludes that most villages "had around twenty houses and probably 100 people or less" (Denevan 1966:58).

Mission populations, by contrast, were much larger—and fluctuated greatly over the period from decimating European epidemic disease. None of the 16 towns in 1713 had more than 2,100 residents, and populations rarely exceeded 4,000 (Denevan 1966:32). A stable food supply was required to support the mission enterprise; its division of labor could not have survived the frequent dispersion of populations of this size. Cattle ensured that stability. Whereas fish (in season) remained part of the diet, wild meat declined (but never disappeared) in importance relative to beef.

Cattle contributed in other ways. Oxen enabled haulage (and still do) and powered sugarcane presses. Beef fat was used for cooking, tallow for making candles and soap. Cowhides, leather goods, and tallow were traded with Santa Cruz and upland cities, as were cotton textiles, cacao, coffee, straw hats, sugar, dried fish, and tamarind seeds (for drinks) (Denevan 1966:32; Vargas Ugarte 1964:73). It is no coincidence that the *vaquero*, or cowboy, enjoyed a status well above that of mere *agricultor*, which was deemed vulgar in the mission hierarchy (Tormo and Tercero 1966:98).

No Spaniard could own land in the province, and *baldíos* (unoccupied, unused land)—which by law belonged to the Crown—were nonexistent, since land not under use at a given time was either flood-prone or considered fallow, and thus essential to maintenance under shifting cultivation. From the Jesuit standpoint, mission land was inviolate and belonged to the Indians (Tormo 1966:117–118). The Jesuits regulated land and resource use between the several towns, working through each one's native *cabildo*, or governing council, a Spanish political form they introduced and that held rights over mission resources (see Garriga 1906:42).

Each family cultivated a plot whose entire harvest, by one account, entered a communal fund controlled by the priests. Garden produce, medicines, clothes, a meat ration, and other items were dispensed fortnightly to each family (Egaña 1966:915; Parejas 1976:39). By another account, the fund took only that part of a family's harvest not needed for its upkeep (Vargas Ugarte 1964:71).

Leandro Tormo (1966:131–133) speaks of a *chacra del cura*, or priest's garden, in each town, owned and worked communally. From it were fed the priests, the several craft masters, town officials, mission guests, the orphaned, the widowed, the aged, and the infirm. The garden further supplied a grain (corn and rice) surplus that was stored for lean times. Cotton and cacao were cultivated in large forest plantations (Denevan 1966:32), which, like the ranches, belonged to the community.

Post-Jesuit Moxos, 1768–1870

Upon Carlos III's expulsion of the Jesuits from Spanish America in 1767, poorly trained missioners (*curas doctrineros*) hastily entered the towns, followed by civil administrators. Thus began long years of decline in the missions as *cura* and administrator whimsically altered Jesuit-wrought forms of land and labor use and vied for control over native labor for personal gain. Unrest marked the early transition period; in some of the towns population declined as Indians returned to the forest (Parejas 1976:64–65).

Church administration of Moxos first fell to Francisco Ramón de Herboso, Bishop of Santa Cruz, who charged his *curas* to attend particularly to the ranches, for "[i]f the Indians lack beef they will begin to hunt monkeys, birds, and whatever else can be taken with arrows, thus living dispersed like infidels, without government or religious doctrine" (Tormo 1966:135–136). The charge was unavailing; cattle were butchered recklessly for quick profits on tallow sales. A governor at the time reported that herd numbers fell from 54,345 to 28,995 in the first 20 years (Block 1980:338).

Such pillage continued during the nineteenth century as large herds of feral cattle, which national governments saw as an unlimited resource, came to roam the *pampas*. Regimes issued letters of credit against Beni cattle and horses, and even paid local officials in cattle (Denevan 1963; Osborne 1964:88). The explorer and railroad engineer

Franz Keller observed in the late 1860s that the government let "adventurers," mostly from Santa Cruz, take cattle from the half-wild herds upon payment of a one-peso head tax to local officials, or *corregidores*. Cattle were butchered only for hides and tallow, which the towns burned as fuel (Keller 1876:182–183).

The mission area planted to cacao and cotton expanded notably until the late 1790s. As cattle numbers declined, cacao and cotton textiles grew in their share of Moxos exports, often at the Indians' expense (Block 1980:338, 342). In San Pedro, where large cacao and cotton plantations preempted nearby forest, natives complained of having to travel long distances to plant food crops. A native official in Trinidad observed that export-crop labor demands there left Indians only one day per week to work their gardens (Block 1980:341).

Once the Jesuits left, Moxos was immediately opened to *cruceño* traders exchanging cloth, steel tools, and factory goods for cacao, cotton, cattle hides, textiles, and forest products. By one account, whites and mestizos of *cruceño* origin made up one-third of the town population in 1850 (Sanabria 1973:19). Then *cochabambinos* entered and, according to René-Moreno (1974:68–69), came to dominate commerce. Keller noted that Indians were at the mercy of unscrupulous traders, which he said the government's four-peso native head tax encouraged (Keller 1875:181–182).

Governments tried at least twice during the nineteenth century to replace Moxos' Jesuit communal structures and ethos with notions of private property and free trade. The Council of the Indies issued an order so directed in 1805, apparently to little effect (see Chávez Suárez 1986:473–474). After Moxos acquired departmental status in 1842, the national government, inspired by liberal doctrines of free trade and economic individualism from industrial Europe, sought to develop the remote region and integrate it into the new republic (see Burton 1978:33–41). To that end, a series of reforms declared the Indians owners of their houses, the land they worked, and the cacao groves they harvested, and provided for taxing them as individuals. Thus was established, albeit tenuously, a jural and administrative structure lasting well into the twentieth century.

The extraction of forest products such as beeswax, vanilla, and especially cinchona bark (for quinine) further livened period commerce. But all pursuits paled beside the Rubber Boom, which had no rival as the most dynamic economic event of the nineteenth century.

From the Rubber Boom to Revolution's Eve, 1871–1948

The rubber era was grim for Beni (and Pando) Indians. By the 1860s, Moxos Indians worked in Brazil's rubber forests (René-Moreno 1974: 402–403). As the boom moved upriver toward Bolivia, the frenzied quest for labor nearly emptied several Moxos towns of men. Company agents, using force and chicanery and aided by local officials, entered the towns and induced natives to sign contracts and go north in river convoys to Acre. Enticements included British silver coins and commodities, which the companies viewed as pay advances and debited to worker accounts. By company design, worker expenditures always exceeded income, creating a permanent debt bond. Debts of deceased workers passed to their survivors.

Bleak conditions surrounded work in the rubber forests; fevers, disease, inanition, and malnutrition took a heavy toll. Extreme unrest rocked some of the Moxos towns. In 1887, natives in Trinidad rebelled in a messianic rising and killed several whites (see René-Moreno 1974:76; Riester 1976:311–312).[2] When authorities savagely responded, rebels fled southwestward across the *pampas* to the towns of San Francisco and San Lorenzo, or into forests southwest of San Ignacio. Many *Trinitarios* living today in the Chimán Forest and the Isiboro-Sécure National Park are descendents of refugees from the Rubber Boom.

José María Urdininea, sent to Beni as prefect in the 1880s to quell the chaos, wrote in about 1887: "This department is completely depopulated owing to the removal, beginning years ago, of its inhabitants to the Madera to collect rubber, from where they rarely return" (quoted in René-Moreno 1974:79). After describing the remaining Indians as slaves, bound to whites by debt peonage, the prefect analyzed the problem and offered a solution:

> These Indians are unable to be citizens because they are incapable of governing themselves; they have the character of children and need the tutelage and protection of the missionary, who to them is both father (*padre*) and defender of their rights.
>
> In a word, to revive Beni it is necessary to protect the Indian, to subject him to the old regime observed by the Jesuits, which produced such brilliant results.
>
> Meanwhile, without the Indian there would be no agriculture, no

river traffic, nothing, because whites here serve only to exploit the Indian (quoted in René-Moreno 1974:80).

With the advent of cheaper plantation rubber in Malaya, the Amazon boom collapsed in 1912, releasing those in its febrile grasp to more equable pursuits. By 1930, on the eve of the Chaco War (1932–35), Beni's socioeconomic order had stabilized. In San Ignacio (see note 1), which received its first influx of white settlers at the close of the boom, about half the native population worked for whites under conditions of debt servitude on cattle ranches and sugar estates, the twin production institutions of the time. The remainder worked their gardens, milled cane on a small scale, and harvested their cacao groves, which all Indians had inherited. Cacao was vital; it was Indian money and was used to acquire necessities.

Most Indians owned at least a few cattle, including a team of all-important draft oxen. Food was plentiful; white employers, or *patrones*, typically fed their workers well, and free Indians either slaughtered their own cattle or bartered for beef. Everybody hunted and fished to supplement the diet. But sinister forces also lurked: an insidious process was afoot that would continue into the 1950s whereby Indians exchanged their cattle, their land, their labor, and their crops, always to their detriment. As one native observer of the time recalled, "The rich brought everything to town and the poor Indians bought. They gave a cow for a hoe, an ax, and a machete."

Bolivia and Paraguay fell to war in 1932, and many Beni peasants fought in the brutal campaigns of the northern Chaco. It was a peasant war, with great carnage on both sides. When peace returned in 1935, Beni *patrones* tried to restore the prewar order on ranches and estates. It was not to be, however; workers were restless and recalcitrant. So began more than a decade of mounting political activity—activity that remained relatively calm in Moxos but that grew turbulent in Bolivia's politically volatile uplands, where it led to the Revolution of 1952.

The Making of a Crisis, 1948–1979

From the late 1940s until 1979 was a dynamic period in Beni. With the Revolution of 1952, the isolated region began to open and link with the uplands. The years mark the beginning of empirical environmental degradation and extensive competition between whites and Indians for natural resources. Three forces dominate the period: expansive cattle ranching, agrarian reform, and an aggressive peltry trade. The interaction of these forces, especially of the first two, had dire consequences for native peoples.

Cattle, some feral, abounded on Beni *pampas* in the late 1940s, numbering by one estimate over a million head in 1952 (Denevan 1963:43). Commercial ranching, pursuant to recommendations of a United States economic mission to Bolivia in the early 1940s, began in 1948 with a project to fly fresh beef to La Paz from Reyes (Clyburn 1970:1–14). When upland beef prices rose sharply after 1950, the airlift, using surplus planes from the Second World War, expanded to supply a lucrative market. Cattle were rounded up for slaughter at rustic airstrip abattoir facilities that began to appear.[3] Many thought the boom ephemeral and butchered indiscriminately in a scramble for quick wealth. Rustling and violence were common, and herd numbers declined by an ominous magnitude, reckoned at between 25 and 65 percent within the first 10 to 15 years of the airlift's inception (Denevan 1963:43; Vivado 1966:9).

Not until 1965 was there some semblance of a cattle industry (Denevan 1963:44). The Federation of Beni Ranchers was founded in Reyes in 1968, and by 1970 Beni was supplying 70 percent of the beef consumed in La Paz and half that consumed in the country (Clyburn 1970:13–14, 19). The state mines, where beef was subsidized, were a critical market. The airlift transformed Beni's relation to the uplands, making Moxos important to the national economy. But one change begot another: as beef prices rose, so did (later, when high prices seemed permanent) Beni land values.

Before the 1953 Agrarian Reform, few Beni landholders held formal titles. Land was plentiful, and the principle of *uti possidetis* governed tenure. Labor had always been scarce, and debt peonage, at least since rubber times, had secured it. Labor was drafted by force and deception, then retained through control of worker accounts and the terms of trade so as to assure perpetual debt. The agrarian reform, whose execution and results in Beni fell short of what they were in the uplands, nullified worker debts and declared peonage illegal, thus striking at the heart of elite production.

But lacking economic alternatives, workers were often unable to enjoy their new freedom. One former *patrón* estimated that 20 percent of San Ignacio's workers remained on estates as day laborers. A tendency to debt peonage continued in labor and trade relations, as McEwen observed in the 1970s in Reyes, where authorities sided with *patrones* who demanded that peasants discharge debts with their labor (McEwen 1975:90).

Workers who quit San Ignacio's ranches and estates either dispersed into surrounding forests to hunt and garden or settled in marginal barrios in Trinidad to farm alluvial canebrakes along the Mamoré. Surviving *patrones* 25 years later, evincing scant sympathy for workers who had by then fallen on hard times, bitterly recalled the banning of peonage, which, they argued, had destroyed agriculture and made Beni a net importer of foodstuffs.[4] The Indian is not able to work effectively without a *patron*, they said.

In San Ignacio, the MNR (*Movimiento Nacionalista Revolucionario*, or Nationalist Revolutionary Movement), the party of the revolution, organized a *sindicato*, or peasant union, naming a white party activist as head. There was talk of making a large garden and working it collectively. But the union never functioned and soon disappeared. The local MNR head (*jefe de comando*), accused by other whites of cattle rustling, quickly acquired and butchered cattle for the airlift; and using native labor, also grew and milled sugarcane. By one account, he supplied estate owners with labor. Indians of San Ignacio in the late 1970s recalled him as an abusive man who kicked and flogged them. Evidence strongly suggests that early MNR officials in San Ignacio, whose post-1952 socioeconomic history is a tragic metaphor of agrarian reform in Beni, used their positions to extort money and favors from people attempting to secure legal title to their land.

That process began later and at first moved slower in Beni and Santa Cruz than in the uplands. By 1966, more than a decade after enactment of the 1953 Agrarian Reform Law, cumulative area distributed in Beni was only 17 percent of that distributed there by early 1975. The corresponding figures for both Cochabamba and Potosí are almost 62 percent, suggesting a titling rate in these upland departments of more than three times that in Beni (or Santa Cruz) over the 13-year period (my calculations using figures from Bolivia 1975:3; and Wilkie 1974:41).

Beef prices continued to climb, and the government announced plans to build roads connecting Beni with Cochabamba, Santa Cruz, and La Paz. Lethargy turned to frenzy. From 1965 until 1975 were years of land rush: land titled between 1969 and 1979 account for 70 percent of all Beni land titled between 1953 and 1979.[5] Whereas cumulative land distribution increased by 40 percent in Cochabamba between 1969 and 1975, it increased by 140 percent in Beni (Wilkie 1974:41; and my calculations using figures from CNRA). Cattle, increasingly the wide-ranging and powerful zebu rather than the traditional "creole," were everywhere in 1979, and few *baldíos* suitable for ranching remained. Beni's "ranching frontier" had closed; only better technology could increase production now (Navia 1988).

The native land problem was acute in San Ignacio de Moxos. In 1976, the Banzer government (see note 10), to placate loyal local elites, issued an executive order (Decreto Supremo 13812) authorizing municipal control over all land within a six-kilometer radius of the plaza, including the power to sell and tax (Bolivia 1976). With the completion in 1975 of a highway between Trinidad and San Borja, through San Ignacio, municipal land values soared. Speculation was rampant, and local control led to enormous graft and exploitation of resident Indians, who lacked the means to secure title, or pay the tax if they did.

Few Indians in San Ignacio in 1978 thus had title to town land where their houses stood, or to ancestral groves (lying within the radius) whose cacao they had long traded for necessities. Whites with title to the groves either cleared them to make paddocks or sold the cacao themselves. Furthermore, much of the municipal land was forested and had long been farmed by the Indians, who were now denied access by white owners with new titles.

From this melee of land rush and reform, Beni Indians emerged landless and destitute in 1979. Government indifference to the problem is remarkable given the Indian fraction of Beni's population. By one estimate, Indian peasants were one-fourth of it in 1979—and 85 percent of Moxos Province (Jones 1980:46). Yet only 0.3 percent of the province's titled land was classed as cultivable in 1975; 98 percent were classed as pasture (Bolivia 1975:15–17; 54–56).

The peltry trade, much of it contraband, thrived in Beni from its inception in the 1940s until the late 1970s. The decline in faunal populations—caymans, peccaries, otters, and felines—has been dramatic. According to local Indians, the large river otter (*londra Pteronura brasiliensis*) was extinct in Moxos Province by 1978, and some caymans (especially *Melanosuchus niger*) were nearly so. Firearms, which Indians first acquired from the sale of jaguar pelts, revolutionized local hunting. Indians had formerly hunted by day with traps, spears, arrows, and clubs, often with dogs. Commercial cayman hunting, by contrast, required combing the lakes and rivers with firearms and flashlights at night. The predation was double on food animals like the peccary, with ominous consequences for the native food supply.

The peltry trade near San Ignacio in the late 1970s was in the hands of five or six itinerant traders, who plied the River Apére and tributaries exchanging consumer goods with riverine Indians. Some Indian hunters, rather than farm, relied entirely on peltry sales to live. The trade turned on credit; every household was indebted, some deeply so. Indians did not reckon trade in money but in peltry units; little cash circulated in the remote upriver areas. As Indians grew poorer, they hunted more, struggling always to reduce mounting debt with peltry.

Expansive ranching and agrarian reform reinforced the peltry trade. Commercial ranching reduced customary native access to beef and dairy products and forced a reliance on game to a degree unknown since pre-Jesuit times. To find game, Indians had to move ever deeper into the forests, where they depended on traders who demanded payment in peltry. Many also retreated into forests remote from the *pampas* to escape marauding cattle, which were ubiquitous by the 1970s and plundered gardens (on high ground) during the yearly floods. Agrarian reform—poorly conceived, underfunded, and riddled with graft—favored ranchers and land speculators to the utter exclusion of Indians.

Economic Malaise, Native Awakening, and the Timber Threat: Official Response and Commitment, 1980–1993

Beni elites' relative affluence in the 1970s gave way to waning fortunes in the 1980s as Bolivia plunged into its worst economic crisis of the century. The decade witnessed happenings of historic moment in Beni: the advent of commercial logging, the return of the Jesuits, and a crystallizing ethnic awareness among the Indians, who began to organize and make demands of the government. On the national front, the government, awed by prospective outside funds and echoing world trends, began to express concern for the environment. And under Bolivia's fledgling democracy, the emerging passions of contending interest groups revolved and pullulated in the public forum.

Bolivia's GNP dropped 17 percent between 1980 and 1984 (Healy 1986:107), and its national debt, a legacy of the Banzer years (see note 10), had reached 4.5 billion dollars by 1985. Black-market cocaine dollars drove the economy by 1984, when annual inflation neared 12,000 percent, the highest in the world (Onis 1989:131). In 1985, the international price of tin, a mainstay of Bolivia's economy since about 1900, fell by half (Crabtree et al. 1987:4)—on the heels of a 30 percent production decline between 1980 and 1984 (Healy 1986:109). As if this were not enough, in 1983 and 1984 one of Bolivia's worst droughts of the century struck parts of the highlands. The production of staples such as potatoes, corn, and barley fell sharply (Healy 1986:109), and entire herds of sheep and camelids perished.[6]

In 1985, spurred by the International Monetary Fund and the World Bank, the government instituted "structural adjustment" measures that included wage freezes, price hikes, public-sector layoffs, currency devaluation, privatization, and an end to price subsidies on food and fuel (Crabtree et al. 1987:14; Newsweek 1989). By the end of 1986, 27,600 workers, or 70 percent of the work force of Comibol, the State-owned mining company that produced 70 percent of Bolivian tin before the crash, had lost their jobs. President Paz Estenssoro declared the tin era over in Bolivia (Crabtree et al. 1987:2, 6–7).

Depressed incomes and layoffs sharply reduced the national demand for beef. Aggregate demand fell further in 1987, when uneconomic exchange rates eliminated the export of live cattle to Brazil. Indeed, contraband cattle from Brazil, Peru, and Argentina reduced Beni's share of a shrinking domestic market (Navia 1988). Severe floods in 1982 killed an estimated 300,000 head, many of them young (Navia 1988). Meanwhile, increasing production costs and tight credit caused a liquidity crisis and forced ranchers, many of whom had entered the decade indebted, to sell cattle. Herd numbers fell by an estimated 20 percent between 1980 and 1988 (Navia 1988), and land values declined. By 1988, ranchers spoke darkly of the end of the cattle era.

Today, cattle are trucked from Beni to Santa Cruz and the uplands over new and improved roads; the airlift continues, but on a small scale. Severe flooding again in 1992, the worst in nearly 50 years, claimed an estimated 80,000 cattle, reducing Beni's herd (2,514,434 head) by three percent (Presencia 1992a). Peasants also suffered steep losses in gardens and houses. Water levels in January of 1993 portended yet heavier flooding and further loss.

Declining land values and cattle numbers brought little relief to native peasants, many of whom by 1988 depended on international food donations. After 1984, destitute uplanders fleeing drought and unemployment entered Beni in growing numbers—from the south through the Chapare and the Isiboro-Sécure National Park, and from the west by road through the Yungas. This invasion further threatened lowland Indians, already driven onto marginal land and into remote forests, and lacking legal titles.

The peltry trade has continued, though evidence hints at a decline. An informed source in Trinidad in 1989 said animals were scarce. A native official in San Ignacio cited weak demand there; peltry had lost value the year before, and peccary were now hunted only for meat. One aggressive peltry trader from the 1970s worked for a timber company in the Chimán Forest in 1989, and was said to be working in cocaine in 1992.

The magnitude and timing of any decline, as its factuality, are moot. The head of the government's Forestry Development Center (CDF, *Centro de Desarrollo Forestal*) reported in 1991 that 37 faunal species (most from the lowlands) faced extinction. Traffic in cayman hides was intense from 1983 to 1988 (over a million hides in 1983), and traffic in live birds and monkeys was brisk during the decade. Between January and July of 1989, CDF confiscated 3,041 collared-peccary skins, 31 white-lipped-peccary skins, 3,559 small-cayman (*lagarto, Caiman yacare*) hides, 236 large-cayman hides, 662 capybara hides, 15 feline pelts, 25 live monkeys, and 7 dead monkeys (Presencia 1989). A decrease in felines had led to a sharp increase in rats near San Joaquín, where residents complained of "bubonic fevers" (Presencia 1991b).

Under pressure from Beni civic leaders, the government placed a three-year ban on hunting in 1986 (ILDIS 1989:100, 200), and CDF reportedly tightened enforcement in 1987 (La Palabra 1988a). But scant resources and Beni's large number of private airfields (see Note 3) for dispatching contraband hindered enforcement (La Palabra 1988b). Any reduction in the peltry trade is thus more likely a result of declining demand, or declining fauna, or both.

The 1980s were a time of ethnic awakening for Beni Indians. In March of 1987, *Trinitario* leaders founded the Center for *Moxeño* Indigenous Councils (*Central de Cabildos Indigenales Moxeños*) in Trinidad. A year later, the Center boasted five affiliates, or *subcentrales*, in south and southwest Beni. Although intended at first to unite *Mojeño* communities (see note 2), which retained active *cabildos* from Jesuit times, the Center later sought to represent all Beni Indians, and to that end changed its name to the Center for Beni Indigenous Peoples (CPIB, *Central de Pueblos Indígenas del Beni*) in November of 1989.

In early 1984, after an absence of 217 years, the Jesuits returned as parish priests to Moxos and southern Yacuma provinces. By late 1988, they had organized eight Christian Grass-Roots Communities. Like the Center, the Jesuits contributed to the ethnic awakening—as did Bolivia's weak democracy; neither the Center's nor the Jesuits' work could have survived the repression of the Banzer years (see note 10). The awakening is also due to events in the Chimán Forest, which has the largest mahogany reserve in Bolivia, maybe in all of South America (La Palabra 1989).

As economic conditions worsened in the 1980s, the central government, pressed by timber interests and by Beni government and civic entities bereft of revenue, opened the Chimán Forest (declared a closed reserve in 1978) in late 1986 to commercial logging, letting concessions to seven Bolivian companies in April of 1987. The concessions covered half (579,000 ha) the forest (Lehm and Navia 1989:309–12). The companies moved swiftly: the *reported* volume (illegal logging is common in Bolivia) of mahogany extracted (excluding Vaca Diez Province) from Beni forests jumped from 8,040 m^3 in 1986 to 19,210 m^3 in 1987 (La Palabra 1988a).

The forest's Indians and loggers have clashed since 1987. By one estimate from a 1988 population survey (La Palabra 1988c), 8,335 Indians (*Trinitarios, Ignacianos, Chimanes, Movímas* and *Yuracarés*), neither consulted nor considered when the forest was "opened," resided either in it or on its fringe (and used its resources). Many had retreated there over the past thirty years and knew it was a last refuge. They had lost the towns, the *pampas*, and nearby forests; there was no place left to go.

In July 1987, six months after the forest was opened and three months after the granting of timber concessions, Conservation International (CI), a private Washington-based organization, brokered the first "debt-for-nature" swap in the Americas.[7] Aided by CDF, CI designed the "integrated" Chimán Program, with socioeconomic, biological, and forestry subprograms. To be implemented by both entities, the program sought to strengthen CDF and to help timber firms log sustainably. In November 1988, the International Tropical Timber Organization (ITTO) approved $1.26 million for the effort over three years. But owing apparently to the program's controversial nature (see Jones 1991a:59), ITTO did not release any funds until 1991.

By mid-1992, most competent environmentalists deemed the Chimán

Program a failure. It was under heavy criticism from respected quarters, as unruly timber firms continued to plunder the forest. CI had become peripheral; a forester sympathetic to the firms managed the program. Management labeled its critics "leftists," and by one account sent deceptive reports to ITTO. LIDEMA (*Liga para la Defensa del Medio Ambiente*, or League for the Defense of the Environment), citing poor subprogram integration and a failure to meet objectives, resigned from the program's board of directors. LIDEMA argued for the program's redesign, while CPIB, noting little concern for conservation or indigenous welfare, argued for its termination. The director of Bolivia's Indigenous Institute (under the agricultural ministry) observed that the program had weak native participation and had not met expectations, yet favored its continuation subject to government "adjustments" (Presencia 1992b).

In 1988, under the Paz Estenssoro government, Indians in the Chimán Forest first demanded a territory there. When they demanded again in 1989, the government formed a commission (funded in part by CI) to study the problem and make recommendations. Without properly consulting with native organizations, the commission proposed that the several ethnic groups receive land on the forest fringe and that the seven timber concessionaires remain in the center. The government left office in mid-1989 without taking action.

In November of 1989, the First Congress of Beni Native Peoples rejected this proposal and demanded a reformulation. The congress also decried the government's indifference since 1988 to demands from natives (mostly *Trinitarios* and *Yuracarés*, threatened by the encroachment of uplanders entering through the Chapare) in the Isiboro-Sécure National Park for a territory there (CPIB 1989). The *Sirionó* did not press for land, near Ibiato, until 1989.

When the new government of Jaime Paz Zamora had not met native demands by late July 1990, Indian leaders gathered near Trinidad to plan a strategy. Among the demands they identified were the right to designate representives before the State, and the right to territories for 10,000 natives in the Chimán Forest, the Isiboro-Sécure National Park, and near Ibiato. During this seminal meeting, the leaders decided in desperation to organize a high-profile march from Beni to La Paz (Presencia 1990a).

On August 15, after a special cathedral mass, 300 Beni Indians, some in native garb and carrying musical instruments, began a 32-day march by dirt road from Trinidad (altitude 236 m) to La Paz (3,632 m), a distance of 650 km over some of the steepest terrain in the Americas (see Jones 1990). The march grew as natives joined it along the way. On September 17, 717 marchers arrived in La Paz, among them 200 women and 55 children (Presencia 1990b).

The odyssey, which ended with another cathedral mass, was epic and moving: impoverished lowland Indians endured dry and dusty roads, searing tropical sun, drenching rains, thin air, and cold Andean nights. They slept in culverts, beneath trucks, on the open ground; they suffered long periods with little water and sparse food. Led by CPIB and self-styled as a march for "territory and dignity," the march was supported by seventy humanitarian organizations, including the Catholic Church and the Bolivian Permanent Assembly on Human Rights (Presencia 1990c).

The government immediately tried to stop the march and discredit its aims. On its eve, and on five occasions during the march, the government made proposals to march leaders in an effort to halt it. Led by the director of the Indigenous Institute, the government sought to divide the Indians, especially to isolate the *Moxeños*, who were pressing for a territory in the center of the Chimán Forest, where the timber concessions were. Bolivia's Vice President, noting (incorrectly) that the Indians were nomads and thus used to walking, said they were going to La Paz for the thrill of it. Furthermore, he said they were landless *because* they were nomads (Los Tiempos 1990).

As the marchers neared La Paz, and as the government appeared to bend to their demands in the face of rising popular support, private sector and civic groups, mostly from Beni and Santa Cruz, rallied to defend *their* interests. Between September 13 and 17, these groups placed seven large ads and open letters in Presencia, the country's leading newspaper. The strongest resistance to native demands came from the *Cámara Forestal* (Chamber of Forestry) and the *Confederación de Empresarios Privados* (Confederation of Businessmen). The ads and letters cited the legality of the timber concessions, the importance of timber royalties (quite low) to Beni's development, the government's credibility with the private sector, and the need to maintain an attractive investment climate.

The enormous popular sympathy for the marchers likely influenced the government's response. On September 24, 1990, Jaime Paz signed four executive orders (*decretos supremos*). Through D.S. 22611, Indians acquired 170,000 ha in the center of the Chimán Forest, an area that came to be called "the multiethnic territory." Firms there, to receive concessions elsewhere in the forest, were to stop cutting by October 31 and to remove all logs by the end of the year. The order also gave the *Chimanes* 400,000 ha in their homeland, along the River Maniqui. Firms there had one year to apply for concessions elsewhere. Indians in the Isiboro-Sécure National Park acquired a million hectares through D.S. 22601. And the Sirionó, through D.S. 22609, acquired 23,140 ha near Ibiato—and an unsolicited 30,000 ha in the San Pablo forest. This last order involved the compensated expropriation of 15 ranches near Ibiato, nine of them with agrarian reform titles, six with titles pending. Within 45 days, government commissions were to begin marking all territorial boundaries.

The fourth order, D.S. 22612, established a commission under the supervision of the Indigenous Institute to draft a lowland Indian law that would confer legal status on traditional leaders and governing bodies, as well as define the relationship between ethnic groups and their territories and resources.

The march, ranked by pundits as a major Bolivian event of the century, forced the lowland Indian land-and-resource issue onto the national agenda and gave it international prominence. Even before the march, the Indian movement and its bid for territories had subtly come to be linked to an emerging national (and international) concern for ecology and the lowland environment. Territories, the argument went, checked environmental degradation. Jaime Paz could thus say in drafting the Earth Letter's Point 22 for the United Nations Conference on Environment and Development held in Rio in June 1992: "Indigenous peoples and their communities, and other local communities, have a vital role in environmental and development matters because of their conditions and traditional practices" (La Razón 1992a).

By executive order (D.S. 22407), Jaime Paz declared on 11 January 1990 a nationwide Historic Ecological Pause—a five-year moratorium on natural-resource exploitation. A bid for time to plan remedial action, the order recognizes Bolivia's growing environmental deterioration, its weak environmental laws and institutions, and its tradition of extractive development models with short-term horizons (Bolivia 1990:5). It stipulates a moratorium on new forest concessions, the rationalization of forest extraction, and improved coordination between State agencies charged with granting concessions and setting population and colonization policies. It also establishes a National Consultative Council for the

Environment to devise operational strategies and plans for sustainable development (Bolivia 1990:7).[8]

On July 11, 1991, Bolivia's congress ratified International Labor Organization Convention 169, thereby making it Bolivian law (No. 1257). Convention 169 (see OIT 1989; Presencia 1990a) recognizes the right of indigenous peoples to participate, through their representative organizations (which the convention legitimates), in all government decisions affecting them, especially in those regarding economic development. It recognizes their rights to a territory and its natural resources, and defines territory as more than land and as including land not occupied but required for subsistence. It sanctions the relocation of indigenous peoples only in exceptional cases, and always with their consent and full knowledge of the reasons.

On April 26, 1992, during an international meeting in La Paz to structure a fund (*Fondo para el Desarrollo Indígena*) to give development aid to indigenous peoples of Latin America and the Caribbean, Jaime Paz issued five more executive orders. Four of them created territories: 115,000 ha (D.S. 23108) for the *Yuquí*, Carrasco Province, Cochabamba; 92,000 ha (D.S. 23110) to the *Araona*, Iturralde Province, La Paz; a territory (D.S. 23111) for the *Chiquitanos*, Santa Cruz; and one (D.S. 23112) for *Mosetenes* and *Chimanes* in the Pailón-Lajas region, on the Beni-La Paz border. The last order also created the Pailón-Lajas Biosphere Reserve (400,000 ha) in the territory (La Razón 1992a).[9]

The fifth order (D.S. 23113) authorized the creation of an Indigenous Forest Guard, to keep unauthorized parties out of the territories and to monitor and protect their natural resources. The order further established that 80 percent of the proceeds from fines and the sale of confiscated equipment would go to Indian communities, and 20 percent to strengthen the Guard (La Razón 1992a). The director of the Indigenous Institute pronounced the Guard a "practical and efficient means" to stem the loss of 200,000 ha of forest yearly to colonists (La Razón 1992a).

The Commitment Questioned

The ground-level impact of these instruments purporting to protect the environment or ensure indigenous welfare has been negligible; current prospects for either look dim. Six months after the march, the government showed little inclination to enforce the executive orders. Popular enthusiasm had waned, and Indian unity had given way to factionalism, even within CPIB, whose leaders politicians now courted. The Catholic Church, under criticism from the Nationalist Democratic Action party (ADN, *Acción Democrática Nacionalista*) and the private sector for supporting the march, was now discreetly divided in Beni: one faction supported the pro-Indian activism of the Jesuits, the other advocated a more moderate role.[10]

In the Chimán Forest, the government did not begin to demarcate the multiethnic territory until July of 1991 (Presencia 1991f). Its boundary, defined in part by rivers of unstable course—and which timber firms sought to keep unresolved—was still in question in Februrary of 1992. In violation of the executive orders, the firms cut and sawed mahogany throughout 1991 (Presencia 1991g), and despite Indian protests, hunted scarce game to feed their workers (Presencia 1991h). In November, Indian leaders complained to Jaime Paz that the government had not kept its commitments and that CDF continued to let cutting contracts (Informativo Andino 1991). In the same month, the government again ordered the companies out by December 31 (Presencia 1991i). The director of the Indigenous Institute said the executive orders had not been enforced earlier because neither he nor CDF had the resources to monitor activities (Presencia 1991h).

In May 1992 Indians in the territory declared an emergency and gave three timber firms until June 21 to leave, saying the government, which they demanded name them forest guards, would otherwise be responsible for the consequences (Presencia 1992c). When the firms ignored the ultimatum, the Indians declared again on September 1 that they would dismantle the mills and use force if they were still there on September 15 (Presencia 1992d).

Throughout 1991 and 1992, woodsmen with power saws posed a major threat to Beni forests. One heard in Beni in 1991 that timber firms supplied Indians with these saws. In the multiethnic territory, 20 or so unidentified men, some armed, were cutting trees indiscriminately in April 1992 (Presencia 1992e). By one informed account, strong capital backed the cutters, whom the timber firms organized and sponsored (Presencia 1992f). The owner of one firm in the Chimán Forest openly declared in February 1992 that he did not hire cutters but did buy logs from them.

Indian leaders denounced the illegal government auction in August

1992, of 2,805 mahogany boles confiscated in the multiethnic territory, saying they would not let them leave (Presencia 1992d). The "auction," conducted with dispatch and with little prior notice, enabled timber firms to acquire the logs at below-market price. Deals were said to have been struck quietly between the firms, the government, and a couple of unauthorized Indians. A year before, firms had acquired another large mahogany lot, mostly from trees cut after the September executive orders and before the year-end deadline for removing them. The firms cut wildly, but heavy rains kept them from meeting the deadline. In a divisive tactic, they struck deals with both CPIB and selected Indians, thus causing disputes.[11] The director of the Indigenous Institute brokered the deals both times, and is said to have received handsome fees.

One *Trinitario* (see note 2) offered a bleak assessment of matters in the Chimán Forest in October 1992. The 1990 executive orders had authorized territories, he said, but the government had done little more; there had been no "consolidation." Timber firms remained in the forest, the Indians were powerless, and some native leaders were now siding with the government (Presencia 1992g).

After the government failed to send the promised boundary commission, the *Sirionó* torched a ranch house in their territory near Ibiato in November 1990. They threatened further burnings unless the government acted (La Palabra 1990a, 1990b). Timely mediation by the Catholic Church apparently averted armed clashes between ranchers and Indians (La Palabra 1990c). When the *Sirionó* began to demarcate their own territory in March 1992, the Indigenous Institute assisted (Presencia 1992h). In June, however, the Indians complained that ranchers remained on 10,250 ha of their 30,000 ha territory; the ranchers wanted compensation, as the government had agreed to in 1990. Noting that the ranchers' agrarian reform titles carried the force of law, whereas a territory carried only the force of an executive order, the *Sirionó* appealed for international help (Presencia 1992i).

Territorial demarcation in the Isiboro-Sécure National Park, never completed, began in October 1991. A band 75 km long and four meters wide along the territory's southern boundary was to be a "red line" against settler incursion (Presencia 1991f). In December 1992, CPIB issued to native officials in half its communities credentials recognizing their authority; those in the other half were to receive them soon, as were officials in the Chimán Forest (Presencia 1992k). The value of the credentials, though, is dubious: In April 1992, a timber firm jailed an official trying to halt illegal logging in the Chimán Forest (La Razón 1992a).

A lowland Indian law, per the relevant 1990 executive order, had been drafted under the Indigenous Institute's supervision and was ready by August 1991 (see Bolivia 1992; Libermann and Godínez 1992:65–73). Despite the law's close adherence to ILO Convention 169 (in which territory is a central concept), CPIB rejected its "intent to give excessive and tutorial powers to the Indigenous Institute" (Presencia 1991e). And persons knowledgeable of jural and Indian affairs cited loopholes that would emasculate the law when challenged in court. The Confederation of Businessmen, backed by Beni ranchers, argued that the law created "mini-republics," and thus violated the constitution, threatened national unity, and retarded development (Presencia 1991c). An anthropologist hired by the Chamber of Forestry described the territorial concept as Marxist, an "experiment that failed in the Soviet Union and in Brazil" (Presencia 1991d).

By December 1992 the government's was one of four pending versions of a lowland Indian law, each prepared by special interests. CPIB and other indigenous entities supported the version prepared by the Indigenous Confederation of Eastern Bolivia (CIDOB, *Confederación Indígena del Oriente Boliviano*).[12] This version, prepared between 1984 and 1992, also embodied Convention 169 (Presencia 1992l). CIDOB's president said the Indians were tired of "slavery" and would not support the businessmen's version, which influential elements in congress favored. The Catholic Church, describing Indians as "the poorest of Bolivia's poor" (Presencia 1992m), also endorsed CIDOB's version; the Archbishopric of Santa Cruz sponsored a petition with 100,000 signatures, which CIDOB delivered to Bolivia's vice-president in early December (Presencia 1992n).

Other new legislation, enacted or pending, also bears on environmental protection and the proposed Indian law. In April 1992, congress passed a new General Environmental Law that provides a broad framework for dealing with a host of issues. Article 5, Point 2, reads: "Promotion of sustainable development with equity and social justice taking into account the cultural diversity of the country." And Article 78, Point 1, reads: "The participation of traditional communities and indigenous peoples in the processes of sustainable development and the rational use of renewable natural resources . . ." (Libermann and Godínez 1992:92, 104). How this law affects (in theory) the environment or native peoples will depend much on its regulatory code (*reglamentación*—in preparation) and on related laws, either pending or soon to pend, and their regulatory codes.

Those related laws include a new General Forest Law and a new Agrarian Law. Privatization, as part of Bolivia's neoliberal reforms, is a key element of both. Privatization is favored by the country's business elites, including the timber companies, which would own large tracts of forest, but is strongly opposed by peasant and environmental groups and by many agrarian and environmental experts (see Urioste 1988, 1992). According to the Bolivian Civil Society Forum on the Environment (FOMODAD, *Foro Boliviano de la Sociedad Civil sobre el Medio Ambiente*), passage of the General Forest Law will have serious environmental and social consequences, including the loss of biodiversity and ecosystems. Passage of both laws will mean "enormous extensions of land and forest in the hands of a few companies and the plunder of rural populations that depend on natural resources" (Presencia 1992o).

Other events further cast doubt on the government's environmental agenda. In November 1992 the press exposed illegal logging in the Amboró National Park (Santa Cruz) by two companies. Inquiries revealed the complicity of CDF officials, including the relative of a senior ADN party boss (see note 10) (Presencia 1992p). When British researchers from the Tropical Agricultural Research Center (CIAT, *Centro de Investigación en Agricultura Tropical*) condemned the action, the agricultural ministry's regional director dubbed them "transformed communists" who "worked only to justify their salaries." The researchers complained that the ministry seemed opposed to CIAT's environmental policies as well as to the defense of the country's legally protected areas (Davies and Johnson 1992).

In November 1992 the Eastern Ecological Association (ASEO, *Asociación Ecológica del Oriente*) denounced illegal logging in the northeast of Noel Kempff Mercado National Park (Santa Cruz) to the park's director, but he took no action. ASEO had denounced logging a year before in the park's southeast, where companies had erected sawmills. Rather than take corrective action that time, the director tried to discredit the informers. ASEO further noted that local politicians (Velasco Province) had made no effort to stop logging or protect the park (Presencia 1992q).

World Wildlife International and Friends of Nature began a long-term research project in 1992 in Santa Cruz's White and Black Rivers

Wildlife Reserve. The project will design a management plan to protect biodiversity under sustained extraction of forest products. Researchers concluded after six months that timber firms there did not log sustainably. They cut trees (mahogany) of less than legal diameter, as well as those with rotten centers (later abandoned) that should have remained as seeders. There was generally little control, with firms operating outside concession areas or in another's concession. Researchers cited the firms' short-term vision and noncompliance with the Ecological Pause and with norms of the International Tropical Timber Organization. And they noted that the Chamber of Forestry's declared aim "to introduce techniques to guarantee sustainable production" did not square with company behavior (Taber and Vallejos 1992).

According to an official of Pando's Peasant Federation in December 1992, the Minister of Agriculture, in violation of the Ecological Pause, had recently authorized timber concessions in Pando (Presencia 1992r), where companies were exploiting a locally severe economic crisis caused by a drop in rubber and Brazil-nut prices to the lowest levels in a century (Presencia 1992s). The official feared intense, uncontrolled logging, with damage to flora and fauna. The concessions included areas previously suggested for wildlife reserves, and the firms were even felling Brazil-nut trees. He asked the government to stop all logging in Pando pending prior "scientific studies." He said the concessions, which peasants did not recognize, would cause violence.

Clearing lowland forest for agriculture also poses a serious environmental threat—one aggravated by corruption. Recent inquiries hint at the extent of official corruption over the past 20 years in the awarding of *Oriente* public land for service or political support. It was rampant during the Banzer and later García Meza military regimes (see Eyzaguirre 1992), and continues under a "democratic" government today.

Recent investigations exposed the involvement of the Minister of Education (an ADN congressional deputy) in a major land scandal in Santa Cruz. BOLIBRAS (Bolivia-Brazil), an enterprise the minister partially owns, received 92,000 ha in the Chiquitanía. Another of his enterprises received 15,000 ha in Ñuflo de Chavez. And his private estate, also in the Chiquitanía, received 4,500 ha. Illegally conveying this public land was the National Agrarian Reform Council (CNRA, *Consejo Nacional de Reforma Agraria*) (Eyzaguirre 1992).

This and other scandals led the government to convene (apparently as propaganda) a commission to investigate irregularities in CNRA and the National Colonization Institute (also empowered to convey land) since 1953, with a "technical audit" to be completed by January 24, 1993 (Presencia 1992t). But because government officials and ADN stalwarts composed the commission, little is expected of its work beyond the implication of a few low-level scapegoats (Aguilar 1992).

Big money now drives land trafficking in Santa Cruz. Public land acquired free in the *Chiquitanía*, whose rich soils and low production costs have attracted Brazilian investors, fetch from $500 to $1,500 U.S. per hectare (Eyzaguirre 1992). A Bolivia Free Movement (MBL, *Movimiento Bolivia Libre*, an opposition party) commission, with the press, visited a BOLIBRAS estate in the *Chiquitanía* in December 1992. Brazilian administrators there told them there was no need for chemical fertilizers, whereas land in Brazil needed 400 kg per hectare. The estate planted wheat in winter, soybeans in summer. Also with the commission was a Federation of Harvesters official who noted that 80,000 workers yearly sold their labor in the lowlands for "a miserable price" because they had no land (Presencia 1992u).

Foreign aid also abets deforestation. The World Bank's Eastern Lowlands Project ($54 million U.S.) is expanding Santa Cruz's agricultural frontier in Ñuflo de Chávez by clearing a million hectares to raise soybeans, sorghum, cotton, sunflower, and other export crops. Heavy equipment cleared 52,000 ha from September to November 1992—by one report, without leaving windbreaks to reduce erosion, as is occurring on nearby Mennonite farms (Presencia 1992v). The government trumpets the social benefits of the project, yet the above-mentioned MBL commission found nearby Indian communities without basic services and complaining that their land had been ravaged. The commission, citing the *Ayoreo* communities of Guidaichay (60 families) and Poza Verde, said the project should commit one million dollars to meet the needs of *Ayoreos* and *Chiquitanos* (Presencia 1992u).

Researchers from Santa Cruz's Tropical Agricultural Research Center (CIAT) point to more than a million hectares of degraded, abandoned farmland in central Santa Cruz. The legacy of a former "agricultural frontier," the land was worked for only 10 years, because of short-term entrepreneurial horizons and resulting poor land management. (Good management includes crop rotation, green manuring, windbreaks, agroforestry, and proper tillage.) The researchers express alarm but not surprise that Brazilians, allied with Bolivians and using the same technologies that degraded Brazilian land, want to expand

Bolivia's agricultural frontier. They ask the Minister of Agriculture not to let this happen (Davies and Johnson 1992).

One must conclude that the government's will, as well as the results of its programs to defend native peoples or the environment, inspire neither confidence nor optimism. In December 1992 the National Coordinating Office for Solidarity with Indigenous Peoples (CNSPI, *Coordinadora Nacional de Solidaridad con los Pueblos Indígenas*) noted the Indigenous Institute's "clear attitude of favoring timbermen and other powerful groups, and its prebendal practice of trying to distort, weaken, and divide native organizations." CNSPI cited the emergence of "'modern' ideologues of racism and colonialism that seek to satanize Indian demands and to delegitimate essential advances like the territories" (Presencia 1992w).

The environmental scene looks little better. A Bolivian economist at the recent Rio Earth Summit noted the country's alarming rate of deforestation and said the Ecological Pause was only a "declaration of intent," devoid of mechanisms to check the destruction. "If the current depredation does not stop, by the year 2000 Bolivian forests will be impoverished, many of them destroyed." (Presencia 1992f).

When a government official declined to provide figures on unpaid timber royalties in Santa Cruz's *Chiquitanía*, saying they were not public, a reporter with figures from civic leaders there said timber firms owed eight million dollars. An official of the Organization for the Defense of *Chiquitano* Social Rights, using volume-shipped figures released by the Chamber of Forestry, calculated the arrears at 36 million dollars. He said the State would need 800 million dollars and 60 years to reforest areas logged recklessly by the firms. And he charged the government with calculated deception—with issuing policies and norms only to create an image that would attract foreign aid (Presencia 1992x).

This charge is percipient, and may explain the government's utter failure to implement and enforce its own environmental policies and regulations—and those regarding the welfare of indigenous peoples as well. In December 1992 Bolivian President Jaime Paz sent a letter to United States President-elect Bill Clinton. The letter, an early bid for continuing economic assistance, reads in part, "The relations between our countries have found a road of constant progress, expressed in common goals: the strengthening of democracy, modernization of the State and society, and the treating of specific topics such as environmental conservation and the struggle against drug trafficking" (Presencia 1992y).

While environmental conservation may be a "common goal," Bolivia's "constant progress" toward it is not evident.

Attitudes, Policy, and "Development": The Backdrop

The historical socioeconomic and political forces depicted above degrade the environment at every turn, often threatening human welfare. Society and environment intertwine in a skein sensitive to events within and beyond the lowlands. A wildly expanding cattle economy in the 1960s and 1970s, for example, forced Indians from traditional land and into the forests, where they survived on wild meat and peltry sales. The international peltry trade drove species toward extinction, thus upsetting the forest equilibrium (animals seed plants) and undermining the native food base.

Likewise, large timber firms initiate a destructive cycle by opening forests with roads to extract high-value woods like mahogany. Smaller firms then enter to extract low-value woods. Then colonists appear, hunting game and turning forest to garden. Colonists have cleared vast areas in Bolivia's *Oriente* since the 1950s. Many entered in the 1980s fleeing drought, unemployment, landlessness, and economic hardship. Also clearing have been agricultural entrepreneurs after quick profits. Perhaps not by chance did Santa Cruz and Beni experience among the worst floods of the century in 1992, with signs of more by early 1993. The water has devastated agriculture and threatens Bolivia's food supply.

With clearing comes burning. Toward the end of the dry season, legions of peasants burn for gardens, and ranchers fire the *pampas* to bring out new grass for now-hungry cattle. In Beni, a blue haze, at times closing airstrips, lingers over the land and dims the sun. In far-away La Paz, the same smoke combines with car emissions to foul the air and irritate the eyes.

Massive clearing has a little-noted sinister potential. Viruses living in faunal hosts are an obscure part of Amazonia's much-acclaimed biodiversity. As forests recede and reduce host habitats, the viruses invade new hosts in their quest to survive—a quest aided by their rapid mutation. By one theory, what became the AIDS virus left receding African forests when a chance mutation enabled it to survive in humans. The prospect of such viruses emerging in Bolivia is frightening, yet real (see Preston 1992).

Further degrading the environment, though not discussed above, are toxic chemicals used to process coca into cocaine (sulphuric acid, ether, acetone), or to refine gold (mercury), now mined in the northeast lowlands and in the Andean foothills. These chemicals pollute ground water and streams, and contaminate fish. It is also likely that the large number of cattle, especially of zebu, on Beni *pampas* and in nearby forest in the late 1970s had (and may still have) pernicious ecological effects. They grazed (and browsed) where they would never have been before 1950.

In all the above cases, degradation has a socioeconomic cause and a socioeconomic consequence. The cause, including ethnic discrimination, derives from time-worn attitudes as well as political and economic policies—which often reflect those attitudes. The attitudes and policies render the Bolivian milieu refractory for pursuing environmental protection or indigenous welfare.

Mining historically has dominated Bolivia's economy, from the colonial era, when silver made Potosí among the world's wealthy cities, to the tin debacle of the 1980s. A "mining mentality," however, has governed the extraction of *all* resources, from metal ores to timber. Planning under this mentality is short term: money must be made now, there may be no chance tomorrow. Bolivia's famed political and economic instability sustains this attitude, which is manifest in rapacious economic behavior. Post-Jesuit Moxos saw administrator, merchant, and cleric pillage mission wealth, from silver church artifacts to growing herds of feral cattle and horses. The Rubber Boom's 30-year frenzy brought fortunes to a few, misery to most. Beni cattle numbers dropped almost beyond recovery in the 1960s as men butchered wildly for quick wealth. Cocaine coca boomed and timber firms plundered lowland forests in the 1980s.

Elite exploitation of Bolivia's Indian majority (about 65 percent of the population) is also a marked historical legacy. Dependence on cheap, native labor is a metaphor of the country's economic history. This racist metaphor runs deep and is part of elite economic culture, especially in the lowlands, where isolation has kept it vibrant. The *Oriente's* cultural attitudes and interethnic relations make it a window on colonial Bolivia.

Strong loyalty to family and a weak State, long a major employer, form the historical backdrop to Bolivia's dominant elite political culture. Since independence, a succession of strongmen have ruled the country, sometimes in the name of a political party, always in an undemocratic political milieu. Until the Revolution of 1952, the Indian masses were excluded from political decision making. Today, a political and economic elite controls their participation by manipulating their leaders and often by the timely (e.g., before elections) and crass disbursement of food and alcohol. Indians rarely participate in the decision-making process except at formal elections, and they do not have equal opportunity or equal access to resources. Today's "low intensity" democracy thus functions very much in a colonial matrix.

Also functioning in that matrix is policy. Many elites view the 1985 neoliberal reforms through a colonial prism that refracts the tenets of free enterprise and deregulation so that they appear to sanction behavior that many nonelites would call depredatory. Further sanction derives from the reforms' endorsement by agencies like the International Monetary Fund and the World Bank. The elite have never felt fettered by law or regulation, or demurred at exploiting social inferiors; they shift "downward" along rigid class and ethnic lines much of the structural adjustment's sacrifice. They meet protest from the underclass by appealing to its patriotism and stressing the sanctity of private property, obedience (ironically) to "economic laws," and the need to attract foreign investment. They cite communism's demise to validate the neoliberal model—which many of them now want to extend to rural areas through proposed new agrarian and forest laws.

This model has deepened what Bolivians call the "social debt." Peasants were poorer in 1990 than in 1985 (World Bank 1990; cited in Urioste 1992:129). It has also rendered the common man's newfound democratic rights less relevant. And while deifying Bolivia's (very) private sector, the model has emasculated and discredited the public sector at a time when the State is needed to deal with complex environmental and other problems.

But of greater concern, the model may be nudging the country toward civil unrest; it has, for instance, hastened rural-urban migration: the 1992 census shows 58 percent of the population as urban, compared with 42 percent in 1976 (Pereira 1993). The cold war's end further augurs unrest. No great-power domestic proxies now oppose each other to contain, albeit imperfectly, rising class and ethnic tensions. In some ways, Bolivia is today as it was in 1940, except that communications have since aroused popular desires for a better life—a life denied the majority.

Related to the neoliberal reforms is rampant corruption—by some accounts, the worst today since independence. The General Secretary of Bolivia's Permanent Assembly on Human Rights said the present government protects corruption and promotes "privileged delinquency" (Sanabria 1993). Sixty-six percent of the respondents in a November 1992 national survey by the Latin American Institute for Social Research (ILDIS, *Instituto Latinoamericano de Investigaciones Sociales*) felt the government had not addressed the problem (Presencia 1992z). And it likely will not, for officials are using the State to secure their future; the current privatization is especially congenial in a setting where public office has commonly been viewed as a chance to bilk the State for oneself and one's family, friends, and supporters.

Likewise embedded in the colonial matrix is agrarian policy. The 1953 agrarian reform spelt great adversity for lowland peasants. It was not well designed for the *Oriente*, where in any case powerful local interests perverted it. And the government, rather than address problems in the uplands, often shifted them to the lowlands, then viewed as fertile and sparsely populated. Santa Cruz was soon a "development pole," and the Chapare, Yapacaní, and Alto Beni became colonization zones. Behind this eastward tilt, and behind foreign immigration as well, was the Bohan Plan, crafted by a 1940s United States economic mission to Bolivia. Among other things, the plan sought to link the *Oriente* to the national economy through capitalist development. Its recommendations guided early efforts of the United States Agency for International Development (USAID) and became part of the MNR's first ten-year plan (Urioste 1992:84, 127).

The agrarian reform also lacked technical support. Agricultural research today, as ever, remains weak—maybe less so in the Chapare, where it enjoys liberal funding from United States antinarcotics programs. USAID, which has much influenced Bolivian development policy since 1952, virtually quit supporting such research in Latin America more than a decade ago. At that time, the United States began to favor the private over the public sector, at home and abroad. Bolivia's public sector has further suffered from its own neoliberal reforms. The country today lacks a research capacity to deal with serious production and marketing problems that bear acutely on the environment: small lowland farmers need alternatives to cutting and burning the forest, while upland farmers need alternatives to migration.

But weak research (coupled with poor extension and no credit) is only part of the dilemma. Population growth, continuing migration, speculation, graft, and an expanding commercial agriculture have all contributed to a land-tenure crisis of sobering dimensions. By 1985, the agrarian reform authority (CNRA) had distributed 32 million ha among 40,000 enterprises, but only 4 million ha among 550,000 peasant households.[13] Considering that on average 1.3 million ha are cultivated yearly, 1 million of them by smallholders, and considering that 4 million ha belong to lowland ranchers, it follows that 28 million ha titled by CNRA or the National Colonization Institute are not in production (Urioste 1992:101). There is more unworked enterprise land today, that is, than there was before the first (1953) agrarian reform. A second reform is needed, it is argued.

These statistics deprive the proposed new agrarian law, drafted by right-wing ideologues with little rural expertise (see Roca and Silva 1989), of its pretended rationale: to increase agricultural productivity. They also challenge a common argument that further production increases require the deforestation of new land. (One wonders what fraction of the unworked enterprise land is permanently degraded.)

It is in the context of a land and resource squeeze that Indians native to the lowlands, numbering from 157,000 (31 groups—Arango 1992:120) to 190,000 (Presencia 1993a), demand territories. Like *Quechuas* and *Aymaras*, lowland Indians lack the notion of land as commodity. Land, unlike, say, an ax, cannot be owned (or sold); it can only be used. The territorial concept permits unrestricted access to land and associated resources for traditional use: hunting, fishing, farming, and gathering. It defines a space for biological, cultural, and economic survival, and for development over the long term (see Chirif et al. 1991). With adequate public support, it can also protect the environment. In Colombia, the 1991 constitution embodies the concept and legislation has made territories communal property. Today, 258 titled *resguardos* occupy 26 million hectares, 20 million of them tropical rain forest (Arango 1992:124–125).

This is not to imply that Indians today are *inherently* benign toward the environment, as they unwittingly tended to be in aboriginal times, when population numbers, native culture, and external forces were in greater harmony. In Bolivia, many Indians have long lived at the impoverished edge of national society, where a crude, potent consumerism has insidiously corrupted traditional values. In the more extreme cases, with their land taken and their labor exploited, they have had to abuse the environment merely to survive. During the 1970s Beni peltry trade, native leaders berated hunters for reckless killing, which they said angered the animals' spirit owners, who sent illness and woe in reprisal. Although these scoldings probably changed little, given the trade's momentum and Indian poverty, they might have done more under less-extreme conditions.

That such pre-Hispanic beliefs still work psychic and social unease after more than three centuries of contact speaks to culture's enduring influence. And indigenous culture, despite any corruption, remains biased toward rational use of the environment, which Indians still have a keen knowledge of—a knowledge they can share with others. Again, with proper support, territories can do much to ensure this environment, with benefits for all.

Further exercising influence on Bolivia's environment and the welfare of its indigenous peoples for many years have been international development agencies. The country's dependence on foreign aid today is extreme, and a growing share of it targets the environment and indigenous welfare. After talks in Washington in September 1992 with Jaime Paz and other Bolivian officials present, the World Bank and several European nations approved $55 million for environmental protection projects, $15 million to be spent in 1993 (La Razón 1992b). USAID recently approved $20 million for the Sustainable Forestry Management Project, to be spent over seven to ten years beginning in 1993 (USAID 1992).

These and future monies will be routed through the National Fund for the Environment (FONAMA, *Fondo Nacional para el Medio Ambiente*), a "decentralized" administrative arm of the presidency established by the new environmental law to collect resources for environmental programs. SENMA's head (see note 8) presides over its board, which includes three representatives from the executive branch, three from departmental environmental councils (created by the new law), and one chosen by private, nonprofit Bolivian agencies with environmental mandates (Libermann and Godínez 1992:106–107). A CIDOB member (see note 12) currently fills this last seat.

Other monies are for indigenous welfare. Announced in December 1992 was a lowland-Indian project designed by the United Nations Development Program (UNDP) and funded by multiple donors at a level of $51 million. The project will support territorial consolidation, environmental protection, education, health, production, marketing, and development decision making (Presencia 1993a). Monies will also enter Bolivia from the La Paz-based Indigenous Development Fund for Latin America and the Caribbean, created in 1992. Jaime Paz proposed the fund at the 1991 summit of Latin American heads of state in Mexico, where it became part of the Declaration of Guadalajara. Inter-American Development Bank (IDB) will act initially as fund trustee, aided by the Andean Development Corporation. Donors will include the World Bank, the Swiss Technical Cooperation, the European Community, the International Fund for Agricultural Development (IFAD), and the governments of Holland, Belgium, and Japan (Presencia 1993b).

These large sums of money have a disturbing effect in Bolivia: persons, usually elites, vie aggressively for key local positions—often lucrative, urban-based administrative positions rather than technical ones—with the government or with donors. Some of these persons are able and committed, but many who are not will be hired (or anyway will be paid) because of family or political ties. Charged with pursuing sound and just environmental and indigenous-welfare goals in the face of graft, racism, and threats to powerful special interests, most of them either will lack power to effect change, or else will not risk their personal peace or socioeconomic standing in Bolivia's small, tight-knit elite society.

Donor projects are too often myopic, poorly designed, or respond to agendas at odds with their spirit and stated goals. The Bohan Plan, for example, in charting a Beni cattle industry, ignored its consequences for a numerous peasant population. Nor was the ommission later corrected: donors supplied credit to Beni ranchers in the 1960s and 1970s (see note 5), yet met the mounting peasant misery—which they knew of—with silence and inaction. During the 1980s, when private-sector support preempted aid, USAID and other donors supplied the Beni Federation of Ranchers with funds for a large, modern, refrigerated abattoir near Trinidad. (The project failed utterly, with the facility today among Bolivian development's "white elephants.")

A careful study of Conservation International's ill-fated role in the Chimán Forest, where it failed to control timber firms, would have value. Poor communication between CI's top management and its Bolivia team, and a desire to be first to broker a debt-for-nature-swap (whose agreement does not mention forest residents), likely figured. But not everybody was blind to the dangers. A year after the swap, a CI forestry consultant observed that one could envision the end of the Chimán Forest if mahogany prices remained high and firms showed little concern for forest recovery. And should this happen, he noted, CI would be blamed (Budowski 1989:346). (The consultant made no mention of native welfare.) Timber companies have since used CI—and ITTO—to seek endorsement for their actions.

The logic of debt relief in exchange for conservation is flawed in Bolivia. This is because the poor bear the debt burden, not the political and economic elite, who contracted the debt in the first place. The elite feel no "relief," only an economic threat. They value debt-for-nature swaps only for the international respectability—and money—swaps attract.

As already noted, the World Bank's Eastern Lowlands Project is problematic for both the environment and Indian welfare. Furthermore, the Bank and CI have favored long-term concessions to encourage timber firms to practice conservation. Indeed, under the proposed new General Forest Law, firms would own the forests. Again, the logic is flawed. The owner of a firm in the Chimán Forest—and today head of the government's Beni Development Corporation (CORDEBENI, *Corporación de Desarrollo del Beni*)—told a reporter that firms would not reforest, since mahogany seedlings took 80 to 100 years to mature and there might not be a market then (Collett 1989). Bolivia's historic instability gives firms a strong present orientation that long-term concessions will not alter. Moreover, an increasingly skewed land and resource distribution threatens domestic peace, and thus future property rights. Timber entrepreneurs know this.

Bolivia's Chamber of Forestry recently cited a World Bank report to argue that logging incidence in Bolivia was low and (per the report) its ecological impact minimal owing to species selectivity (Presencia 1992za), but in 1988 concessions accounted for 40 percent of the total forest area (Davies and Johnson 1992), and the figure has since risen. Such reckless, narrowly focused reports cause great harm to the environment and to human welfare.

USAID's strong influence on Bolivian development since 1952 has been noted. Consonant with America's Bolivia policy, antinarcotics has preempted USAID's agenda since about 1978. What was then "coca substitution" is today "alternative development." But owing to its antidrug bias, this "development" is narrowly conceived, and has in any case been secondary to "interdiction," which aims to reduce coca processing through the use of Bolivian and American police and military force. Discord between interdictors and developers has at times strained relations within the American Mission. In 1991, USAID, apparently afraid to offend SUBDESAL's director and the Embassy, demurred at confronting a growing and harmful politicization of the Bolivian agency executing United States-funded alternative development projects (see Jones 1991b).

Disconcerting evidence suggests that the United States is sacrificing "development" to local elites in return for their support of interdiction—and now of a new extradition treaty. If so, the drug war has replaced the cold war in Bolivia, with development again secondary to other agendas—and with USAID's Sustainable Forestry Management Project an empty and costly tribute to the trendiness of the times.

Admonition for Donors

This chapter argues that international aid often fails to protect Bolivia's lowland environment or promote native welfare. Rather, by funding an elite-controlled apparatus little committed to either, it subsidizes degradation and racism. The voluminous aid now entering Bolivia invites waste and duplicity. Reliable sources say the government spent nine million dollars on its 1992 Seville exposition, and cite frequent and lavish junkets to Spain by officials and partisans. Donors must not be duped by the government's Ecological Pause, or its unctuous declarations of support for Indians—all designed to attract money, all promoted by a pricey public-relations effort, including lofty and well-timed pronouncements, pretentious high-level meetings, and glossy brochures prettied with flashy pictures.

Donors should press the government hard to protect the environment and promote indigenous welfare; in Bolivia, the two cannot be divorced. The hour beckons: The Nobel Peace Prize recently went to a Guatemalan Indian; the United Nations has declared 1993 the International Year of Indigenous Peoples; a government promising to protect the environment has taken office in the United States; and Bolivia will hold national elections in 1993. A few simple admonishments are thus in order regarding development aid.

First, development agencies need to design their projects better. Designs tend to reflect elite interests, to the detriment of indigenous and marginal groups. Agency staff need to understand better the relations between economic, social, and political forces in the countries *and regions* where projects operate. Too often, they make neither the time nor the effort to understand. To include Bolivians on design teams is not enough; rather, teams must include representatives of Indian and marginal groups potentially affected by projects. Agencies often accept persons, even Indians, supplied by the government, which usually has little interest in addressing native welfare.

Related to this admonishment is a second: Development agencies should be independent of bilateral foreign policy. Or perhaps better said, that policy should respond to other than short-term or narrowly conceived self-interest. The welfare of Bolivia's indigenous majority is in the long-term best interest of the United States and other donors. Cold war policies toward Latin America have particularly harmed United States interests (e.g., illegal aliens), by protecting privilege and suppressing ethnic, class, and regional demands for justice for fifty years. About four centuries maturing, those demands today pose a major challenge to democracy, development, and peace in the Americas.

Cold war policies are dated; a new era calls for new policies and new international structures. The long-sacred tenet of national sovereignty is weakening; on the horizon are glints of an order where ethnic, regional, and class entities play a greater role. As a practical start, donors might press Bolivia to institute a civil service scheme, and might make aid contingent upon concrete progress toward eliminating racism and protecting the environment. Donor concerns for individual human rights might be expanded to embrace collective rights.

Third, development agencies should better monitor their projects. The above-cited 1992 ILDIS survey showed the Catholic Church and the media as the democratic institutions Bolivians most trust (Presencia 1992z). Project-funded monitoring units, independent of the Bolivian government and of donor agencies, should exist and should include Church and media personnel as well as those from affected marginal groups. The units should report directly to the *original* funding source—in the case of USAID projects, probably to the General Accounting Office, United States Congress.

Fourth, donor governments, assisted by donor domestic groups with expertise in environmental and indigenous affairs, should better monitor public funds going to such multilateral agencies as the World Bank.

Fifth, there should be major *external* evaluations of development projects about every two years. These should be commissioned by *other* than donor agencies, or at least by other than their offices disbursing funds in Bolivia, who have a vested interest in project performance and often pressure evaluators to bias reporting.

A Closing Word: Early 1993

As 1993 begins, the drama continues. In early January, Harvard's Jeffrey Sachs visited Bolivia as a government guest. This was the first of several consulting visits planned for the distinguished economist over three months under a grant from the Andean Development Corporation; ADC's president is the government's former planning minister. Sachs will assess the impact of the neoliberal reforms instituted by the MNR in 1985 and will offer advice on addressing current problems (Presencia 1993c). He met with notables from the public and private sectors, including the head of the Confederation of Businessmen and ADN leader Hugo Banzer (see note 10), the government's candidate for president in the June 1993 elections (Presencia 1993d).

Sachs observed that although the government's economic management since 1985 had curbed public spending, checked inflation, and brought stability, it had not paid the "social debt." He noted that living conditions were unequal between the poor, stagnant rural sector and the modern, urban, free-market one, and said that the two sectors had to be integrated (Presencia 1993d). He noted the low internal investment rate (10 percent of GNP) and stressed the need to invest in agriculture, which makes up half the economy (La Razón 1993): "There cannot be 10 percent growth when the rural half of the country lacks access to credit or any form of investment, and has no land titles. . . . A single country cannot have two societies and two economies" (Presencia 1993d). One might add that this state of affairs existed in 1985—or, for that matter, in 1885.[14]

Again, for elites the colonial prism refracted Sachs's remarks. The president of the Confederation of Businessmen said the "social problem" was a "time bomb" that could destroy the model, leverage "populist currents," and erode "modernization." After seven years, the model had been implemented only "half way"; more effort had to be made to include Bolivia's majority in the free market so they could enjoy its benefits (Presencia 1993e). Editorials in *La Razón*, a conservative La Paz daily, said the implications of Sachs's remarks were clear: the model should be extended to the countryside through passage of proposed new agrarian and forest laws.

Bolivia is at a crossroads in 1993. The cold war's end has profoundly altered a half-century-old order. In the United States, a liberal Democratic government has just entered office, after 12 years of conservatism.

Bolivians, aware that the neoliberal reforms have favored the country's elite and deepened social ills, will elect a president in June. The interplay of international and domestic forces, as always, will affect the environment and Indian welfare.

The MNR's candidate in the June elections is Gonzalo ("Goni") Sánchez de Lozada, a mining magnate who as planning minister in 1985 initiated the neoliberal reforms. His running mate is Víctor Hugo Cárdenas, a noted *Aymara* educator and intellectual, and leader of the Tupac Katari Revolutionary Liberation Movement (MRTKL, *Movimiento Revolucionario Tupac Katari de Liberación*). For the first time in Bolivia, an Indian is running for high office on a major party slate. Cárdenas, a unique and able man who has come a great social distance, has publicly condemned timber firms for abusing Indians (see Cárdenas 1990), and also opposed the spirit and content of many of the neoliberal reforms. His candidacy is ironic—and risky for the MNR; some elites refer to him as Goni's *pongo*, or Indian servant.

Idealists read into the slate the MNR's desire to break free of a colonial past and continue what it began in 1952. Realists cite Bolivia's history as a dark and sober reminder that the slate may merely reflect another elite bid for power and for control of the Indian populace. As these words are written, preelection government paralysis and political posturing grip this improbable land of intrigue, contradiction, and irony, a land where things are rarely what they seem—and a land where time may be running out.

NOTES

The final version of this chapter, completed in early 1993, describes events and institutional relations up to that time. The discussion mentions national elections held in May 1993 as forthcoming, elections that resulted in the election of Gonzalo Sánchez de Lozada and Víctor Hugo Cárdenas as president and vice-president of Bolivia. Upon taking office in August 1993, the new administration embarked on a major reorganization of state agencies. As a result, the institutional actors described in this chapter have changed substantially. The historical processes that threaten the livelihoods of lowland Native People have not.

1. Much of the material here is based on a study of Indians in and near San Ignacio de Moxos that I conducted while living there for two years, between 1976 and 1979. Many of these Indians today live in and around the Chimán Forest. I returned to Beni briefly each year from 1988 to 1992. Like Bolivians

with some historic sense, I often use the term "Moxos" to refer to Beni, whose limits approximate those of the Jesuit province. San Ignacio is the capital of one of Beni's eight provinces, which bears the same name. The Inter-American Foundation, the National Science Foundation (BNS-7709610), and the University of Florida Foundation Tropical South America Program funded my research in the 1970s.

2. I follow local custom and refer to non-Indians as whites (*blancos*). The term's referent, however, is cultural. Descendants of mission Indians usually refer to themselves by the names of the mission towns—*Trinitarios* from Trinidad, *Ignacianos* from San Ignacio. Descendants of Indians from Santa Ana de Yacuma, however, call themselves *Movímas*. The term *Moxeño* refers to Arawak speakers descended from Indians in missions in southern and southwestern Beni. Most *Moxeños*, with an estimated (1991) population of 30,000 (Arango 1992:120), are *Trinitarios* or *Ignacianos*.

3. There were 35 private airstrip slaughtering complexes in 1965. By 1975, Beni boasted 700 airfields, most of them on ranches (Nagashiro 1975:84). (The number of private airfields was estimated at above 4,000 a decade later.) Their presence was not lost upon drug traffickers in the 1980s—or upon ranchers who had by then fallen on hard times.

4. Beni was self-sufficient in food production in 1930. In the late 1970s, by contrast, sugar sold in Trinidad and San Ignacio was produced in Santa Cruz, and a decade later, sugar and many other items came from Brazil, through Guayaramerín.

5. The economist Ronald Clark remarked the "high rate of development and new settlement" during the 1960s, especially after 1968. He attributed this to a quadrupling of beef prices over a five-year period and to increases in the number and amount of foreign loans to the Agricultural Credit Bank for cattle ranching (Clark 1974:31–32). Also, the rate of land distribution increased generally in Bolivia beginning in 1968 with the creation of "mobile brigades" to expedite the process. The brigades operated in Beni until about 1975; the sharp increase in area titled after 1968 in some degree is owed to their presence (see Wilkie 1974:35).

6. But there was a "boom within the crisis": the cocaine trade thrived (see Healy 1986). By one estimate, one in six Bolivian families earned a livelihood from coca and cocaine by 1985 (Crabtree et al. 1987:84). Several elite families from Beni and Santa Cruz profited enormously. Indeed, the crisis favored the boom, as drought forced many desperate highlanders into the Chapare to work coca to survive.

7. CI bought (via Citicorp Investment) $650,000 of Bolivian debt from private holders for $100,000, donated by the Frank Weeden Foundation. The agreement annuls the debt in return for the creation of a $250,000 endowment ($100,000 from the Bolivian treasury, $150,000 from USAID's PL-480 program) in local currency, to be administered by the Bolivian Academy of Science and the Ministry of Agriculture, and used to manage and protect the Beni Biosphere Reserve. The Bolivian government would protect the reserve and establish three

adjacent buffer zones: the Chimán Sustainable Production Forest (670,000 ha), CORDEBENI Watershed Protection Zone (225,000 ha), and the Yacuma Regional Park (130,000 ha). CI would provide technical and administrative support (Dogse and von Droste 1990:29–30).

8. The Consultative Council originally operated under SEGMA (*Subsecretaría General del Medio Ambiente y Recursos Naturales*, or General Subsecretariat for Natural Resources and the Environment), in the Ministry of Agriculture. But the new General Environmental Law (April 1992) eliminated SEGMA and placed all environmental matters under SENMA (*Secretaría Nacional del Medio Ambiente*—National Secretariat for the Environment), which operates directly under the presidency.

9. CI played a constructive role by ushering the measure through the Bolivian bureaucracy and financing required socioeconomic studies (see Rioja Ballivián 1992).

10. ADN is Santa Cruz-based and led by Hugo Banzer, *cruceño* military president (favored by Washington) from 1971 to 1978, when Beni and Santa Cruz elites prospered as others suffered brutal repression and severe economic hardship. The same lowland families that supported Banzer in the 1970s support ADN today. ADN articulates timber and ranching interests and controls economic affairs in the country's "patriotic accord" ruling coalition dominated by ADN and MIR (*Movimiento de la Izquierda Revolucionaria*, or Movement of the Revolutionary Left). The coalition has taken MIR far to the right of its populist origins. Bolivians often associate ADN with the drug trade. "Not all ADN supporters are drug traffickers, but all drug traffickers are ADN supporters," a saying goes. Circumstantial evidence also links Banzer to the trade (see Dunkerley 1984:315, 318, 319).

11. The 1990 executive order regarding the Chimán Forest is often read as giving the Indians rights to logs remaining after 31 December 1990, but the issue is legally murky, as may also be that of rights to timber felled illegally in native territories. The Indigenous Institute's director said in May 1992 that 10,000 board feet had been confiscated in the Isiboro-Sécure National Park and would be auctioned off, with half the proceeds going to Indians there (Presencia 1992j). CPIB leaders were disturbed by the 1992 auction of Chimán Forest mahogany, which they had planned to sell at market price for money to build schools and improve communities. They received instead 549,440 Bs. (US $140,000) at the time of the auction. By December 1992 the government had frozen CPIB's account after the Grand Chimán Council (see note 12) complained that *Chimanes* had not been allotted a share (personal communication).

12. Founded about 1982, CIDOB is the oldest multiethnic lowland indigenous organization. Although its name suggests a broader reach, CIDOB's influence goes little beyond Santa Cruz. (No truly pan-ethnic organization exists.) CPIB, founded in 1987, represents groups in central and southern Beni. The Grand Chimán Council (*Gran Consejo Chimán*) represents *Chimanes* near San Borja, western Beni. New Tribes Mission, long active there, urged its founding in 1989, apparently to check CPIB, which NTM saw as a Catholic-inspired rival. The

Indigenous Center for the Amazon Region of Bolivia (CIRABO—*Central Indígena de la Región Amazónica de Bolivia*), representing groups from northern Beni, La Paz, and Pando, emerged in about 1990, encouraged by the Indigenous Institute's director, a MIR activisit from nearby Riberalta. Public and private entities manipulate and divide these groups in pursuit of their own interests.

13. In Santa Cruz, about 9 million ha have been distributed, 28 percent to smallholders, who account for 96 percent of all titled units. Medium and large holders account for 4 percent, but own 72 percent of the land. In Beni, 10.4 million ha have been distributed: 0.4 percent to small producers, who account for 57 percent of all units; and 99.6 percent of the land to medium and large holders, who account for 43 percent of the units. In Pando, 1.7 million ha have been distributed, only 0.9 percent to smallholders. Bolivia's surface area measures 108 million ha (Urioste 1992:103).

14. Sachs was in Bolivia as ADN guest when the reforms were instituted in 1985, and is often associated with them, but according to the finance minister at the time, Bolivians alone designed them. Sachs, after first expressing doubts, endorsed the reforms some days later, "as if he had proposed them," in a memorandum to Hugo Banzer (Cariaga 1993). The issue of his 1985 role aside, one wonders what Sachs, who has since advised Poland, Russia, and other countries, knows about Bolivia, where the long-held notion of Indians as cheap labor breeds idleness among the elite and mingles with their present-time orientation and their marked propensity to consume (also historic) to favor a low investment rate.

REFERENCES

Aguilar Dávalos, Yuri

1992 La Explotación Maderera y el Centro de Desarrollo Forestal. Presencia. 24 Dec. La Paz.

Arango Ochoa, Raúl

1992 Derechos Indígenas sobre el Territorio. *In* Territorio y Dignadad: Pueblos Indígenas y Medio Ambiente en Bolivia. Kitula Libermann and Armando Godinez, eds. 118–129. La Paz: Instituto Latinoamericano de Investigaciones Sociales (ILDIS) and Editorial Nueva Sociedad.

Block, David

1980 In Search of El Dorado: Spanish Entry into Moxos, a Tropical Frontier, 1550–1767. Ph.D. dissertation, Ann Arbor, Mich.: University Microfilms International.

Bolivia, Government of

1975 El Proceso de Reforma Agraria en Cifras. La Paz: Consejo Nacional de Reforma Agraria.

1976 Decreto Supremo No. 13812. Gaceta Oficial de Bolivia 17 (868): 25.044–25.049.

1990 Pausa Ecológica Histórica. La Paz: Presidencia de la República; Ministerio de Asuntos Campesinos y Agropcuarios; Subsecretaría de Recursos Naturales Renovables y Medioambiente. República de Bolivia.
1991 Proyecto Ley de los Pueblos Indígenas del Oriente, el Chaco y la Amazonía. Presidencia de la República, Ministerio de Asuntos Campesinos y Agropecuarios. La Paz, Bolivia.
Budowski, Gerardo
1989 Informe sobre la Visita al Campo en el Bosque de Chimanes, 23 Junio–2 Julio, 1988. *In* Nuesto Bosque de Mañana. La Paz, Bolivia: Instituto Latinoamericano de Investigaciones Sociales (ILDIS).
Burton Rodríguez, Guillermo
1978 Departamento del Beni: Su Creación. Trinidad: Universidad Boliviana "José Ballivián." Mimeo.
Cárdenas, Victor Hugo
1990 El Bolsillo de los Madereros o la Vida de los Pueblos Indígenas. Presencia. 1 Sept. La Paz.
Cariaga, Juan L.
1993 Sachs y lo Que Pueden Hacer los Bolivianos por sí Mismos. La Razón 24 Jan. La Paz.
Chávez Suárez, José
1986 Historia de Moxos. La Paz: Editorial Don Bosco.
Chirif, Alberto, Pedro García, and Richard Smith
1991 El Indígena y su Territorio Son Uno Solo. Oxfam America and Coordinadora de las Organizaciones Indígenas de la Cuenca Amazónica (COICA). Lima: Epigrafe.
Clark, Ronald James
1974 Land-Holding Structure and Land Conflicts in Bolivia's Lowland Cattle Regions. Inter-American Economic Affairs 28(2): 15–38.
Clyburn, Lloyd E.
1970 U.S. Assistance to Bolivia in Agricultural Development. La Paz: United States Agency for International Development. Typescript.
Cochrane, Thomas T.
1973 Potencial Agrícola del Uso de la Tierra de Bolivia. Publicación de la División de Recursos de Tierra, Administración de Desarrollo de Ultramar, Gran Bretaña. La Paz: Editorial Don Bosco.
Collett, Merril
1989 Bolivia Blazes Trail . . . to Where? Christian Science Monitor. 10 July.
CPIB (Central de Pueblos Indígenas del Beni)
1989 Conclusiones del 1er Congreso de Pueblos Indígenas del Beni. 10–13 Nov. Trinidad: CIDDEBENI (Centro de Investigación y Documentación para el Dessarrollo del Beni). Photocopy.

Crabtree, John, Gavan Duffy, and Jenny Pearce
1987 The Great Tin Crash: Bolivia and the World Tin Market. London: Latin America Bureau.
Davies, Penny, and James Johnson
1992 Reservas Forestales y Parques Nacionales. Presencia. 29 Nov.
Denevan, William M.
1963 Cattle Ranching in the Mojos Savannas of Northeastern Bolivia. Yearbook of the Association of Pacific Coast Geographers 25:37–44.
1966 The Aboriginal Cultural Geography of the Llanos de Mojos of Bolivia. Ibero-Americana 48. Berkeley and Los Angeles: University of California Press.
Dogse, Peter, and Bernd von Droste
1990 Debt-For-Nature Exchanges and Biosphere Reserves: Experiences and Potential. Paris: UNESCO.
Dunkerley, James
1984 Rebellion in the Veins. London: Verso Editions.
Egaña, Antonio de
1966 Historia de la Iglesia en la América Española. Desde el Descubrimiento hasta Comienzos del Siglo XIX. Madrid: Biblioteca de Autores Cristianos.
Eyzaguirre, Gloria
1992 Parlamentario Interpela al Estado a Nombre de Tres Millones de Campesinos. Presencia 29 Nov. La Paz.
Garriga, Antonio
1906 Linderos de los Pueblos de las Misiones de Mojos, Declarados y Confirmados por el Padre Provincial Antonio Garriga en su Visita de 10 de Octubre de 1715. *In* Jucio de Límites Entre el Perú y Bolivia, vol. 10. Victor M. Maurtua, ed. 34–42. Madrid: Imprenta de los Hijos de M.G. Hernández.
Healy, Kevin
1986 The Boom within the Crisis: Some Recent Effects of Foreign Cocaine Markets on Bolivian Rural Society and Economy. *In* Coca and Cocaine. Cultural Survival Report No. 23. Deborah Pacini and Christine Franquemont, eds. 101–143. Cambridge, Mass.: Cultural Survival.
ILDIS (Instituto Latinoamericano de Investigacines Sociales)
1989 Nuestro Bosque de Mañana: Sintesis Documental del Proceso Forestal Beniano: 1979–1988. La Paz: ILDIS.
Informativo Andino
1991 Pueblos Indígenas Intentan Proteger sus Territorios. 10 December. Lima, Peru.
Jones, James C.
1980 Conflict between Whites and Indians on the Llanos de Moxos, Beni Department: A Case Study in Development from the Cattle

Regions of the Bolivian Oriente. Ann Arbor, Mich.: University Microfilms International.
1990 A Native Movement and March in Eastern Bolivia: Rationale and Response. Institute for Development Anthropology Network 8 (2):1–8.
1991a Economics, Political Power, and Ethnic Conflict on a Changing Frontier: Notes from the Beni Department, Eastern Bolivia. Working Paper No. 58. Binghamton, N.Y.: Institute for Development Anthropology.
1991b Institutional Analysis of the Programa de Desarrollo Alternativo Regional (PDAR). Report prepared for USAID/Bolivia. Binghamton, N.Y.: Institute for Development Anthropology.

Keller, Franz
1875 The Amazon and Madeira Rivers: Sketches and Descriptions from the Note-Book of an Explorer. Philadelphia: J. B. Lippincott and Co.

La Palabra (Trinidad, Bolivia)
1988 Actividades Desarrolladas y Proyecciones del CDF-RN 1 (38) (15 April).
1988b CDF-RN No Autorizó Ninguna Exportación de Pieles 2 (68) (6 Sept.).
1988c Subcentral de Cabildos Ignacianos Envió al MACA Resultados del Censo Poblacional del Area de Chimanes 2 (84) (7 Nov.).
1989 Este Año, se Controlará Explotación en Chimanes 2 (99) (10 Jan.).
1990a Pueblo Sirionó Invadirá Hoy Estancia Ganadera. 25 Nov.
1990b Indígenas Sirionós Incendiaron Estancia Ganadera. 27 Nov.
1990c Ganaderos y Pueblo Sirionó Arriban a Acuerdo de Paz. 2 Dec.

La Razón (La Paz, Bolivia)
1992a Los Cuidadores del Mundo. 26 April.
1992b Programas Ambientales Contarán con $US 15 Millones en 1993. 4. Oct.
1993 Bolivia Debe Ser Paciente y Persistente para Atraer Inversiones. 9 Jan.

Lehm, Zulema, and Carlos Navia
1989 Conflictos Sociales en el Bosque de Chimanes. *In* Nuestro Bosque de Mañana: Sintesis Documental del Proceso Forestal Beniano: 1979–1988. 309–324. La Paz: ILDIS.

Libermann, Kitula, and Armando Godínez, eds.
1992 Territorio y Dignadad: Pueblos Indígenas y Medio Ambiente en Bolivia. La Paz: Instituto Latinoamericano de Investigaciones Sociales (ILDIS-Bolivia), and Caracas: Editorial Nueva Sociedad.

Los Tiempos (Cochabamba, Bolivia)
1990 Para el Vicepresidente de la República Marcha Indígena es "Buena Aventura." 4 Sept.

McEwen, William J.
1975 Changing Rural Society: A Study of Communities in Bolivia. London: Oxford University Press.

Métraux, Alfred
1942 The Native Tribes of Eastern Bolivia and Western Matto Grosso. Bureau of American Ethnology Bulletin 134. Washington, D.C.: United States Government Printing Office.

Muñoz Reyes, Jorge
1977 Geografía de Bolivia. La Paz: Editorial Don Bosco.

Nagashiro Ribera, Gustavo
1975 Geografía. *In* Beni, Pando y Tarija. Monografía de Bolivia, Vol. 4. 71–90. La Paz: Biblioteca del Sesquicentenario de la República.

Navia Ribera, Carlos
1988 Ganadería Beniana en Desastre. La Palabra 2 (81) (25 Oct.) Trinidad, Bolivia.

Newsweek
1989 Perestroika Goes South. Newsweek. 6 Nov.

OIT (Organización Internacional de Trabajo—International Labor Organization)
1989 Convenio No. 169 Sobre Pueblos Indígenas y Tribales en Paises Independientes. Lima, Peru: Oficina Regional de la OIT para America Latina y el Caribe.

Onis, Juan de
1989 Brazil on the Tightrope toward Democracy. Foreign Affairs 68 (4): 127–143.

Osborne, Harold
1964 A Land Divided. London and New York: Oxford University Press.

Parejas Moreno, Alcides
1976 Historia de Moxos y Chiquitos a Fines del Siglo XVIII. La Paz: Instituto Boliviano de Cultura.

Pereira M., René
1993 La Pobreza, la Principal Partera de la Urbanización. La Razon (La Paz) (13 Jan.).

Presencia (La Paz, Bolivia)
1989 CDF Decomisó Cueros. 2 Sept.
1990a Pueblos Indígenas Piden al Gobierno Reconozca a Sus Autoridades. 11 Aug.
1990b Marchistas Indígenas Llegan Hoy a La Paz. 17 Sept.
1990c Indígenas del Beni Inician Marcha a La Paz. 15 Aug.
1991a Indígenas de Bolivia Ganaron Autonomía de Sus Propios Territorios. 18 July.
1991b 37 Especies Animales Silvestres en Proceso de Extinción. 24 Oct.
1991c Ley Indígena Conformará "Republiquetas" Dentro del País y Frenará el Desarrollo. 25 Sept.
1991d Antropólogo Afirma que Ley Indígena Creará "32 Republiquetas" Etnicas. 27 Oct.
1991e CPIB Rechaza Papel "Tutor" del Estado en Comunidades. 28 Sept.
1991f Comienza Demarcación de Parque Isiboro Sécure. 17 Oct.

1991g Indígenas de Territorio Multiétnico Denuncian Saqueo de Madera. 22 Nov.
1991h Empresas Madereras Exterminan a los Animales Silvestres. 25 Oct.
1991i Madereros Conminados a Salir de Territorio "Multiétnico." 23 Nov.
1992a Por Inundaciones se Perdió Mas de 91 Mil Cabezas de Ganado. 17 Mar.
1992b Según Gobierno, Programa Chimanes no Colmó Expectativas. 31 July.
1992c Indígenas dan Plazo a Madereras para que Saquen sus Motosierras. 31 May.
1992d Indígenas de Territorio Multiétnico Analizarán Propuesta del CDF. 1 Sept.
1992e Reducen Volumen de Corte de Madera en Bosque Chimán. 12 April.
1992f Bolivia Perderá la Riqueza de Sus Bosques hasta el Año 2000. 13 June.
1992g Indígenas Pedirán a Juan Pablo II Ayuda para Consolidar Territorios. 10 Oct.
1992h Sirionós Reforestan su Territorio. 17 March.
1992i Sirionós Esperan Apoyo Internacional para Recuperar Territorios. 11 June.
1992j Guardia Indígena Forestal Comenzará a Decomisar Madera. 24 May.
1992k Autoridades Indígenas Tienen sus Propias Credenciales. 21 Dec.
1992l Nueva Ley Indígena Contribuirá a Crear un Nuevo Estado Nacional. 25 Oct.
1992m Se Insta a Buscar Consenso para Compatibilizar la Ley Indígena. 29 Nov.
1992n Vicepresidencia no Cumplió su Palabra sobre Ley Indígena. 15 Dec.
1992o Bosques y Tierras en la Mira de la Privatización. 29 Nov.
1992p Establecido Nuevo Puesto de Control en el Parque Amboró. 21 Dec.
1992q Director del Parque Noel Kempff Conocía Tala Ilegal y no la Denunció. 19 Dec.
1992r Gobierno Autorizó Explotación de Madera en Pando. 29 Dec.
1992s 1992: Las Mejores Realizaciones y Peores Frustraciones de Pando. 3 Jan.
1992t Gobierno no Privatizará la Tierra ni Formulará Nueva Ley Agraria. 30 Dec.
1992u Existe Penetración Brasileña en Tierras Pretendidas por BOLIBRAS. 24 Dec.
1992v Tierras Bajas del Este son Desboscadas Masivamente. 22 Nov.
1992w Critican a IIB por Intentar Manipulación de Indígenas. 15 Dec.

1992x Autoridades Chiquitanas Exigen Pago por Regalías Madereras. 18 Dec.
1992y Jaime Paz Confia en Política Antidroga de Bill Clinton. 29 Dec.
1992z Iglesia Mantiene Preferencias de Confianza en la Población. 18 Dec.
1992za Según Cámara Nacional Forestal, BM Recomienda Política Forestal con Aprovechamiento de Bosques. 31 Aug.
1993a Se Determinará Territorio y Mejorará Producción en Poblaciones Indígenas. 9 Jan.
1993b Cancilleres de Iberoamerica Buscarán Fortalecer Fondo Indígena. 11 Jan.
1993c El 21060 no Fue Relocalizado pero Debe Pagar Deuda Social. 8 Jan.
1993d No se Puede Crecer 10% Cuando la Mitad Rural no Tiene Nada. 9 Jan.
1993e Pobreza es una Bomba de Tiempo Que Podría Hacer Fracasar el Modelo. 10 Jan.

Preston, Richard
1992 Crisis in the Hot Zone. The New Yorker. 26 Oct.

René-Moreno, Gabriel
1974 Catálogo del Archivo de Mojos y Chiquitos. La Paz: Librería Editorial Juventud.

Riester, Jurgen
1976 En Busca de la Loma Santa. La Paz and Cochabamba: Editorial Los Amigos del Libro.

Rioja Ballivián, Guillermo
1992 Conquista del Pueblo Chimane, Pilón Lajas, Territorio Indígena. Presencia (La Paz). 10 May.

Roca, José Luis, and Oscar Silva León
1989 La Necesidad de una Nueva Ley Agraria. Proyecto de Ley Agraria. La Paz: Ministerio sin Cartera. República de Bolivia.

Sanabria Fernandez, Hernando
1973 En Busca de El Dorado. La Paz: Librería Editorial Juventud.

Sanabria, Floren
1993 Corrupción e Impunidad. Presencia. 8 Jan.

Taber, Andrew, and Cristian Vallejos
1992 Actividad Maderera y Vida Silvestre. Presencia. 27 Dec.

Tormo Sanz, Leandro
1966 El Sistema Comunalista Indiana en la Región Comunera de Mojos-Chiquitos, Part 1. Comunidades (Spain) 1 (1):96–140.

Tormo Sanz, Leandro, and Javier Tercero
1966 El Sistema Comunalista Indiana en la Región Comunera de Mojos-Chiquitos, Part 2. Comunidades (Spain) 1 (2):89–117.

Urioste F. de C., Miguel
1988 Segunda Reforma Agraria. No. 1. 2d ed. La Paz: Centro de Estudios para el Desarrollo Laboral y Agrario (CEDLA).

1992 Fortalecer las Comunidades: Una Utopía Subversiva, Democrática . . . y Posible. La Paz: AIPE/PROCOM/Tierra.

USAID (United States Agency for International Development)
1992 Sustainable Forestry Management (511–621). Project Identification Document. La Paz.

Vargas Ugarte, Rubén
1964 Historia de la Compañia de Jesús en el Perú. Vol. 3. Burgos, Spain: Aldecoa.

Vivado P., Manuel
1966 Beef Cattle in the Eastern Sub-Tropics, Bolivia 1965: Pre-Feasibility Study. Vol. 1. La Paz: Ministerio de Asuntos Campesinos y Agropecuarios. Mimeo.

Wilkie, James W.
1974 Measuring Land Reform. Los Angeles, Calif.: UCLA Latin American Center.

World Bank
1990 Informe sobre el Desarrollo Mundial 1990: La Pobreza. Washington, D.C.: World Bank.

Chapter 6

The Social and Economic Causes of Deforestation in the Peruvian Amazon Basin: Natives and Colonists

Eduardo Bedoya Garland

The accelerated deforestation presently occurring in the Amazon basin has inspired literature from multiple perspectives ranging from the forest sciences and ecology to the social sciences. This literature, diverse as it is, offers no consensus on the criteria we should use to evaluate a problem of such magnitude. Some authors attempt to explain deforestation as a function of the technological characteristics of swidden agriculture (Myers 1980). Others seek the structural causes in the impact of state economic policies on the Amazonian region (Hecht 1984), in models of capital accumulation applied to each region and state (Schmink and Wood 1987; Collins 1986; Collins and Painter 1986), in the unequal rates of exchange between country and city (Painter 1987), in territorial conflicts (Foweraker 1981; Martine 1980), or in the propagation of inadequate technologies for peasant agriculturalists by state agencies (Todaro 1977). Some academics focus their analysis on specific institutional aspects of the problem, such as the legal system of land tenure or the government colonization programs that enable settlers to relocate to the ecological and demographic frontiers (Moran 1984).

In this chapter we demonstrate that the natural resources of various production systems are managed in specific ways that in turn influence the rate of deforestation. An analysis of the *colonos* (colonists) and indigenous groups of the Amazon reveals that the rate of deforestation by colonists is greatly affected by a high level of market integration; their relative access to land, labor, and capital; and the formal land tenure system. For the Peruvian indigenous groups, the significant factors are the degree of access to land, the size of the productive unit, and the availability of family labor.

The discussion is divided into four parts. The first briefly describes the agricultural expansion in the Peruvian Amazon Basin and its relation to deforestation. The second discusses the relationship between intensification, deforestation, and the limits of the institutional changes brought about by the deeding of land titles; here we determine the impact of coca expansion and its relation to deforestation and agricultural intensification in the Upper Huallaga (the principal coca producing region in the world), using information from a 1981 survey of settlers. Part Three compares settler production systems with those of three native groups, and conclusions are presented in Part Four.

Agricultural Expansion and the Family Economy in the Upper Peruvian Rain Forest

As a result of increasing migration by highland peasants and the expansion of urban demand for foodstuffs, the Amazon Basin—particularly the upper rain forest—has become the second most productive agricultural zone in Peru, and in some respects the most important. Data collected by the National Survey of Rural Families in 1984 (INE 1987) indicates that hard yellow corn and rice harvested in the Amazon region account for 67.5 percent and 48.3 percent, respectively, of the total national acreage planted in these species. Sixty percent of the family farming units in this region produce corn, while 71.2 percent plant rice.

When we analyze the yields per hectare of these crops in the Amazon, however, we find that corn yields are 20 percent below the national average and rice is 32 percent below average (Verdera 1984: 181–183). Almost all of the yellow corn and two-thirds of the rice in the Amazon is cultivated using slash-and-burn agriculture. Slash-and-burn techniques rely on crop rotation and lengthy fallow periods, utilize planting cycles that correspond to the rainy seasons, and incorporate little in the way of modern technology (INE 1987). Only 10 percent of the family units in the rain forest actually use modern inputs (ibid.). Irrigated rice has average yields below those for the Pacific coast because irrigation techniques in the Amazon are still rudimentary.

The ever-increasing importance of tropical agriculture in the overall national economy comes at great ecological cost. An average of 350,000 hectares per year are cleared for shifting agriculture. Sustained utilization of this land is extremely low. Given current trends, only 20 percent of the total 11 million deforested hectares in the Peruvian Amazon will actually be in use for agricultural or animal production by the year 2000, while the remaining 80 percent will be in fallow. Even though slash-and-burn is diminishing because of the diffusion of more stable and intensive types of soil management, up to 50 percent of the land in some valleys, such as that of the Central Huallaga, is still cultivated under this system (Dourojeanni 1982). A better understanding of the causes of these ecological and migrational phenomena requires a special analysis of the settlers who continue to practice shifting agriculture and who deforest the region at steadily increasing rates.

Sixty-one percent of the family farming units in the rain forest lack legal title to the land they work (INE 1987). Although slash-and-burn agriculture demands continuous migration, which influences and complicates the process of land deeding, the existence of such a high percentage of farmers without legal title to their land has social, economic, and political origins.

It is in the context of the low agricultural productivity of this region and a prolonged national economic crisis that we see, especially in the Upper Huallaga valley, an increase in the response to the illicit demand for cocaine that expanded in the United States and Western Europe from the mid-1970s through most of the 1980s. Coca production in this region has soared from less than 5,000 hectares in 1975 to at least 60,000 in 1986 (Bedoya 1987; ECONSULT 1986). Approximately $840 million per year is circulated in the upper rain forest from the processing of coca leaf into cocaine paste and its later sale (ECONSULT 1986). This sum represented 35.4 percent of the legal agricultural national product of Peru for 1985 ($2.372 billion), a percentage that has undoubtedly not diminished in the past six years.

Dourojeanni (1989) indicates that approximately 200,000 hectares in Peru are planted in coca, that coca production has resulted in the deforestation of nearly 700,000 hectares since the 1970s, when the number of coca producers began to increase, and that coca is responsible for 10 percent of the total deforestation of the Peruvian Amazon in this century. The figure of 700,000 hectares includes land planted in coca, land planted in subsistence crops to support the coca producers, land that has been abandoned as a result of decreasing coca yields, and land cleared for the construction of landing strips necessary for the transport of the coca paste (Dourojeanni 1989). Dourojeanni also notes that the increasing number of coca producers has indirectly increased deforestation by pushing farmers growing crops for legal markets farther and farther away from the zones of terrorism and drug trafficking. The program of manual eradication aimed at reducing the amount of coca production also aggravates the destruction of the rain forest (Bedoya 1990). When peasants in the Upper Huallaga region learned of the eradication program, they moved to more remote areas of the valley or to such regions as the Central Huallaga and Aguaytía zone of Pucallpa. The dynamics surrounding the coca eradication program have led to a general dispersion of coca plantations throughout the entire tropical rain forest.

Because the cultigen is illegal, two-thirds of the coca producers in the Department of San Martín are settled outside the boundaries of the Special Project of the Upper Huallaga. The Project's boundaries were created to implement a program for cultivating substitutes for coca production (see fig. 1). Coca producers, both inside and outside the boundaries, have chosen locations remote from the main highways in order to avoid police repression. One such site, the Tingo María region of the Upper Huallaga Valley, is officially considered a National Park. Such other National Parks as Abiseo in the Department of San Martín, and such National Forests as Von Humboldt in Ucayali and Huánuco, or Biabo in San Martín and Ucayali, have also been invaded by coca producing peasants. If the economic crisis in which Peru currently finds itself continues to worsen, if the terms of exchange between city and countryside remain unfavorable to farmers, and if the international demand for cocaine continues to increase, the expansion of coca production will persist. If the market shifts in favor of another drug, however, coca production will cease to be a tool for dealing with the economic crisis, or at least it will become a less important one.

The export of cocaine paste parallels in many respects the export of rubber and other tropical products—such as—coffee and cocoa, except that this processed commodity is not legal. Rasnake and Painter (1989) argue that the impact of cocaine export on social life—on the mobilization of labor, the distribution of wealth, and the creation of enclaves—is analogous to the impact of legal export industries that have dominated Bolivia's national economy. For example, export commodity production often relies on labor migrants as a work force. Indeed, coffee and coca in the Upper Huallaga have shown many of the same relations of produc-

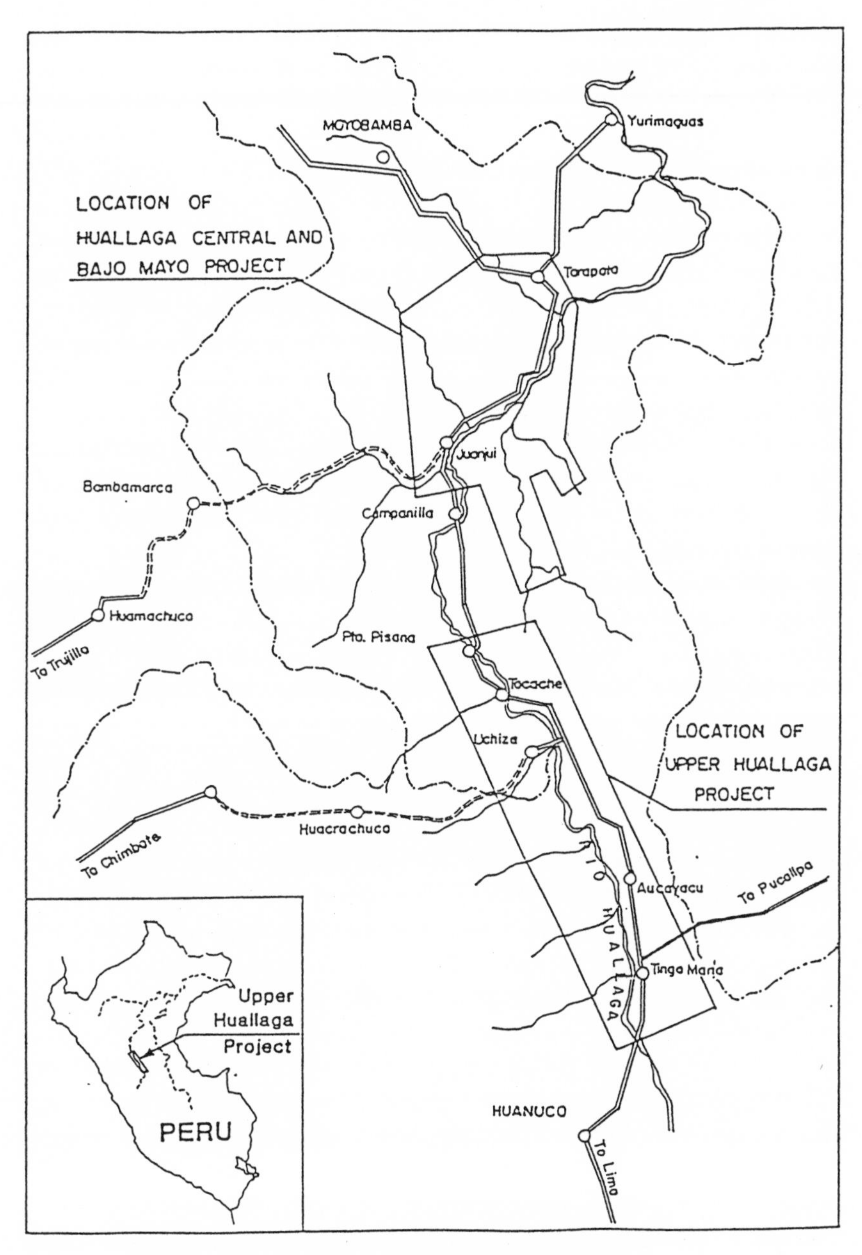

Fig. 1. Huallaga Valley. (Data from Survey FDN 1981.)

tion, including temporary labor force and relatively high salaries. In other words, coca production for the international market is a replacement for coffee, cocoa, lumber, and rubber production.

Colonization and Shifting Agriculture

Although colonists were pulled by the market into the tropical zones for local cash crops (Durham 1977), one of the primary reasons underlying the migratory movements is the continuing impoverishment of peasant populations of the Andean highlands. In 1973, an important study of migratory colonists in the Tingo María, Campanilla, and Tocache regions found that 42 percent of the settlers in this zone had migrated because of acute shortages of land in the highlands, while another 26 percent had moved because of the lack of available work. In other words, 68 percent of colonists in this region relocated for explicitly economic reasons (CENCIRA 1973).

These economic forces shape the perceptions and expectations that Andean farmers bring to the region. The highland peasants view the rain forest as a place where they will always be able to satiate their "hunger for land." Shoemaker summarizes the attitude of migrants from the sierra to the Satipo settlement area:

> What the peasant migrant sees in the mountain is, in a word, land. They come from areas in the sierra, or highlands, where they might have possessed a few plots of land that altogether would not add up to more than one or two hectares. The opportunity to acquire 20 hectares, the average size of a tenant farmer's land holding in the forest, is a strong incentive to move. Looking back, they always say that the rich people in their original communities only owned five hectares, therefore 20 hectares appears to be practically a state. (1981:96)

In the highland population's concept of the rain forest as a zone of limitless potential and inexhaustible land, however, the fragility of the tropical ecosystem is rarely considered. The real availability of land suited for agriculture is significantly less than the initial perceptions of the Andean colonists. Within this context of land scarcity in the Andean zone and the illusion of an abundance of land in the Amazon region, it is not surprising that settlers utilize extensive agriculture as the main form of soil management in the tropical lowlands.

Even apart from the fact that this perception of an abundance of land is false, it is an insufficient explanation for the Andean farmer's use of extensive agriculture as a farming technique. When a settler from the highlands decides to cultivate a plot in the Amazon basin, he is beginning a new phase in his life. After four to six years, he is able to send for his family, with the goal of reproducing the traditional scheme of the peasant economy using unremunerated work. The migrants begin to raise such food as corn, manioc, and bananas, most of them annual crops, with the dual objectives of providing for their own needs and selling the remainder. One of the major errors in the various settlement programs has been the initial assumption that agriculture in settlement areas would be based largely on intensive capitalist farming and that production would be organized in terms of profit maximization. Several studies indicate that during the initial phases of resettlement most peasants are unable to organize their production in this manner (CENCIRA 1973; Bedoya 1981). In the early years, the pioneer settlers lack basic resources and must therefore grow crops for their own consumption. State agencies do not usually lend money for producing subsistence crops (Schuurman 1980). The lack of available capital for buying such necessary inputs as fertilizers encourages extensive agriculture as a way to harvest enough food for family survival. At any rate, in the upper rain forest as a whole, colonists rely on crop rotation and fallow periods to regenerate their land during the initial phases of settlement.

The lack of money further impedes agricultural intensification, by restricting the colonists' ability to contract wage labor. Labor scarcity is one of the most serious difficulties the colonists confront. Because settler families recently arrived in the tropical zone are young and thus have few members, their greatest constraint is labor. According to a 1981 survey among the colonists of Alto Huallaga, more than half of the families with less than four years in the area had problems of insufficient labor, while for those families who had been in the zone for eight years or longer the problem was relatively insignificant (FDN 1981). Because recent immigrants prefer working their own land before hiring themselves out to neighbors and other farmers, the opportunity cost of colonists in recently settled areas is quite high. In other words, labor scarcity is a function of the size of the family, the abundance of land, and the price of labor. In the Tingo María region during the 1950s, 1960s, and the first half of the 1970s, there was intense competition for labor between family farms, on the one hand, and large coffee and tea plantations, on the other (Bedoya 1981).

The two most profitable crops during this period were coffee for export and tea for domestic consumption. This meant increased labor scarcity for the colonists, as they were not able to pay equivalent salaries for work on crops that were not as lucrative.

All of these factors have implications for the use of natural resources, in particular the use of extensive agriculture in the Upper Huallaga region. To the extent that new land continues to be available and labor and money scarce, the lowland colonists prefer to open new plots rather than regenerate the soils with modern inputs. Furthermore, weeding is extremely laborious. Field surveys and estimates by the state agencies indicate that the amount of labor needed to clear virgin forest is less than that needed for weeding, which tends to increase labor demand as the years pass. For example, a study on coffee growing in Tingo María in 1962, a period of relative land abundance, indicated that from the second year on, a parcel of land planted in coffee, bananas, and manioc required 175 percent more man-days for weeding than the initial land clearing of the virgin forest (SIPA 1962: 114).

Moreover, since higher crop yields do not necessarily guarantee greater future income, the settlers cannot justify spending scarce financial resources on modern inputs. During the initial stages of resettlement in the frontier region, therefore, colonists combine a strategy of reducing the requirements for labor with that of minimizing risk. Such an approach leads to a pattern of extensive land management and continuing deforestation.

Availability of Land, Intensification, and Deforestation

The Foundation for National Development (FDN, Fundación para el Desarrollo Nacional) reported in 1981 on the family economy of 352 agriculturalists living in the Upper Huallaga region in an area delimited by the Upper Huallaga Special Project. The principal characteristic of this region is that although the area in perennial crops is slightly greater than that actually cultivated in annual crops—52 percent and 48 percent, respectively—fallow agriculture in 1981 covered the bulk of land in the interior of the Upper Huallaga Special Project, and 79 percent of the farmers had holdings whose total area in annual crops plus fallow took up more space than did permanent crops. Furthermore, 75 percent of the colonists who predominantly used the fallow land system (which is in reality a migratory agriculture) had holdings that were larger than 15 hectares, while 55 percent of the colonists who practiced migratory agriculture without fallow had holdings of less than 15 hectares.

FDN's estimate that around 50 percent of the colonists used modern inputs is misleading. The farmers of the region may use improved seed, fungicides, pesticides, or fertilizers, but almost never all simultaneously. In the context of relative land availability, the use of modern inputs seems to be a last-minute decision. After having covered his labor needs, the colonist might decide to buy fertilizers or some other input (Bedoya 1987). Generally, in 1981 the area in the Upper Huallaga Special Project was characterized by extensive land use. The larger the holding the more extensive the means of cultivation. Economic orientation and lower population density in some parts of Upper Huallaga affected the decision to use land extensively (Bedoya 1987; Aramburú and Bedoya 1987).

The practice of extensive agriculture in this region is also significantly one of the impacts of coca production on legal agriculture in Upper Huallaga. Land, capital, and labor have been directed since the 1980s to coca production, more profitable than any legal crop. In the interior of the Upper Huallaga Special Project such crops as rice and yellow maize, cultivated using a fallow system, have been jeopardized by increasing production costs brought about by the expansion of coca cultivation. For example, the wages paid by coca producers are at least 125 percent higher than the salaries paid by legitimate farmers. In order to compete with coca producers, legitimate farmers must increase the wages, but they can do it only for short periods of time, and the rest of the agricultural tasks remain uncovered. Statistics referring to land use in 1986 show that this tendency towards extensive land use continues (ECONSULT 1986). The intensification of land use was concentrated above all in the areas outside the boundaries of the Special Project, where the greatest number of monocrop coca producers exist.

FDN's report also permits us to determine accurately how the rate of deforestation (number of hectares under permanent crops + annual crops + fallow grass divided by the number of years a plot is occupied) correlates with the total amount of land available per colonist. In table 1 we see that the rate of deforestation by agriculturalists who utilize the soils least intensively—0.1 to 0.2—is 1.88 hectares per year.[1] This is in contrast to areas that receive more intensive use—0.7 to 1.0—where the annual rate of deforestation involves 0.97 hectares, almost half that produced by lower intensity use. In general terms, we can see from table 1 that the higher the intensity of soil use, the lower the rate of deforestation. If we consider intensity as a variable that corresponds primarily to plot size, we can formulate a relationship between the size of the agricultural unit and the rate of deforestation.

Table 1 also illustrates the surprising amount of damage caused to the forest by the low intensity usage of soils. As shown in the table, peasants must clear 18.93 hectares of land in order to provide 3.52 hectares for cultivation at a level of intensity of 0.1 to 0.2. At levels of 0.3 to 0.4, a total of 15.16 hectares must be cleared in order to cultivate 6.19 hectares. In the former case, the proportion of the effective area to be cultivated to the total area deforested (fallow + cultivated areas) is 5.37. In other words, those following a course of extensive soil management must clear five hectares of land for every one hectare in production. These data show the enormous amount of forest land being consumed as a direct result of extensive agricultural land use practices. Moreover, in the most extensive patterns of soil use, the annual increase in the number of hectares in fallow is higher than the annual increase in the number of cultivated hectares (see table 1).

The newest settlement areas of the upper rain forest have low population densities and relatively available land in comparison to the Andean highlands. This relative abundance of land has promoted a return to less intensive agricultural systems in the tropical areas, in contrast to the more intensive farming systems characteristic of the highlands. Such a response is a technological regression, a phenomenon most commonly associated with agricultural and demographic frontiers (Boserup 1984).

In every valley of the Peruvian upper rain forest except that of Alto Mayo, the more recently settled areas tend to develop extensive patterns of cultivation and have lower population densities (Aramburú and Bedoya 1987: 160–161; Bedoya 1987). In the Upper Huallaga region, the area most recently settled and with the lowest population density is Tocache-Uchiza where in 1981 approximately 68 percent of the plots had land-use intensities oscillating between 0.1 and 0.2, and where around two-thirds of the colonists did not use modern inputs. In contrast, in the Tingo María region, the area settled the longest and with the highest population density, 66 percent of the plots had land use intensities between 0.6 and 1.0, and the majority of the farmers used modern inputs, although very unsystematically (Aramburú and Bedoya 1987). The Upper Huallaga case confirms Boserup's hypothesis, but we must qualify

TABLE 1. Annual Rate of Deforestation and Annual Increase of Agricultural and Fallow Hectares, by Intensity of Soil Use in Alto Huallaga, 1981

Intensity of Soil Use[a]	No. of Family Units	(1) Average Size of Plots (Lectares)	(2) Average No. of Cultivated Hectares	(3) Average No. of Fallow Hectares	(4) Total Agricultural Area (2 + 3)	(5) Average No. of Years Occupying the Plot	(6) Annual Rate of Deforestation (4 ÷ 5)	(7) Annual Increase of Agricultural Hectares (2 ÷ 5)	(8) Annual Increase of Fallow Hectares (3 ÷ 5)	Annual Rate of Deforestation (4 ÷ 5) or (7 + 8)
0.1–0.2	86	36.94	3.52	15.41	18.93	10.05	1.88	0.35	1.53	1.88
0.3–0.4	87	26.40	6.19	8.97	15.16	10.31	1.47	0.60	0.87	1.47
0.5–0.6	59	20.82	7.45	5.14	12.59	9.43	1.34	0.79	0.55	1.34
0.7–1.0	54	15.89	8.02	1.84	9.86	10.15	0.97	0.79	0.18	0.97

Source: Data from FDN 1981.

[a]Intensity of soil use $= \frac{\text{(annual crops + permanent crops)}}{\text{(annual crops + permanent crops + fallow area)}}$

the argument. In contrast to the colonists from the Upper Mayo region, the farmers from Upper Huallaga had migrated from the Andean provinces of Huánuco without any previous experience in such tropical intensive crops as irrigated rice, and with less disposable capital (Bedoya 1987). In Upper Huallaga the extensive use of land in the recently settled areas is a consequence not only of lower population pressure, but also of the technical background of the colonists and their relatively limited resources.

Land Tenure, Deforestation, and Intensification

Table 2 relates the system of land tenure to the deforestation rate. The deforestation rate among squatters is 2.72 hectares annually, a figure more than double the rate of deforestation on the land belonging to legal tenants. This fact suggests that the system of land tenure also influences the rate of deforestation, but other explanations must be considered. First, squatters have higher rates of deforestation because they have only recently begun to cultivate their land, in comparison to legal tenants or grantees who have been farming their plots for more years (see table 2). Second, squatters clear the forest at a higher rate because they cultivate predominantly annual crops (Bedoya 1987). Of course, it is the insecurity of tenure that leads the squatters to cultivate annual rather than perennial crops. In Upper Huallaga, the deforestation rate diminishes among those agriculturalists with legal title to land only when these families begin to focus on the production of more permanent crops, which are strictly commercial.

As table 2 illustrates, the index of intensity of soil use is almost the same for both legal tenants and squatters. In other words, agriculturalists possessing titles to their property rely on forms of extensive soil management. The deeding of property in the Upper Huallaga has not resulted in more intensive use of land because the expanding coca-producing plantations in the valley create a chronic labor shortage. The majority of the agriculturalists who have title to their land do not manage coca plantations. As we have mentioned, the coca producers have the profits to be able to pay daily workers substantially higher wages than legitimate farmers in the region can afford (Bedoya and Verdera 1987). Without a reliable work force, the settlers are unable to intensify agricultural production. The banks, in authorizing credit for legal producers, fail to take into account the high costs of labor generated by the

TABLE 2. Annual Rate of Deforestation and Annual Increase of Agricultural and Fallow Hectares, by Land Tenure in Alto Huallaga, 1981

Land Tenure	No. of Family Units	(1) Average Size of Plots	(2) Average No. of Cultivated Hectares	(3) Average No. of Fallow Hectares	(4) Total Agricultural Area (2 + 3)	(5) Intensity of Soil Use[a]	(6) Average No. of Years Occupying the Plot	(7) Annual Rate of Deforestation (4 ÷ 6)	(8) Annual Increase of Agricultural Hectares (2 ÷ 6)	(9) Annual Increase of Fallow Hectares (3 ÷ 6)	Annual Rate of Deforestation (4 ÷ 6) or (8 + 9)
Legal tenants	187	25.87	6.8	9.39	16.19	0.42	12.61	1.28	0.54	0.74	1.28
Precarious	155	24.57	5.84	7.74	13.58	0.43	7.49	1.81	0.78	1.03	1.81
without legal title											
By purchase	125	24.33	5.70	7.25	12.95	0.44	8.03	1.61	0.71	0.90	1.61
Squatters	30	25.61	5.72	8.58	14.3	0.40	5.25	2.72	1.09	1.63	2.72

Source: Data from FDN (1981).

[a]Intensity of soil use $= \frac{\text{annual crops + permanent crops}}{\text{annual crops + permanent crops + fallow area}}$

coca crop and underestimate the costs of seasonal labor. Within the context of such an unfavorable cost structure, the possibilities of increasing and intensifying legal agricultural production are minimal.

The expansion of the coca economy in the region has also powerfully deterred the development of technology specifically related to the production of legal crops, and thus has impeded the intensification of agriculture and thereby increased deforestation. Our field experience and the data found in the FDN study (FDN 1981) indicate that the annual rate of deforestation from coca production is less than that of holdings cultivated predominantly in annuals. The problem of deforestation caused by coca production does not seem to be as serious when it is analyzed plot by plot, until we look at the large number of farmers, both inside and outside the borders of the Upper Huallaga Special Project, who have cleared steep slopes to cultivate coca.

Structural factors that directly affect the level of intensity of soil use and therefore also the rate of deforestation are plot size, the market price of different crops, the cost of modern inputs, and other variables such as regional population densities. In Upper Huallaga a simple change in the legal system is no guarantee that a change in the pattern of soil use will follow. Finally, because the Upper Huallaga region is profoundly influenced by the expansion of coca production, we cannot generalize from it to all of the upper rain forest region. Without the profound price distortions that are found in coca production, changes in the legal land tenure system probably could modify soil use intensity.

Our focus on family economies should not give the erroneous impression that Upper Huallaga is an aggregate of traditional peasant units. The income of a coca producer can be as much as ten times that of a dry rice farmer, and there is a notable difference in income between a colonist who has recently migrated to Upper Huallaga and whose crops are mostly annuals, and a long-term producer of coca, cafe, cacao, or tea. Moreover, among the coca producers themselves there is a clear economic gap between the recently migrated colonists who plant coca from the first year of clearing a plot—now a fairly common phenomenon because of the national economic crisis—and those coca producers who came to the region in the fifties and sixties. Years of association with the narcotics trade have allowed the latter to attain a certain level of capital accumulation and annual income.

Differentiation between family units occurs in this way: after a certain number of years, a group of colonists, often families with a larger number of working members, starts to cultivate cash crops that are favored by higher national and international prices and bring in greater revenues. The transition from a pioneer peasant economy to simple commodity production—or even to petty bourgeois production, using Friedmann's (1980) classification—represents a significant step toward higher incomes and a greater articulation with the market economy. The colonists in the early stages tend to orient their survival strategies according to a peasant economy, while in the later stages they tend to orient their economic strategies according to petty bourgeois production. In the early stage, agricultural activities are basically covered by family labor, and even though most of the produce is sold on the market, subsistence consumption may reach 35 to 40 percent of the harvest. In the later stage, wage labor constitutes the principal relation of production, and produce is almost totally sold on the market (Aramburú and Bedoya 1987).

Not all Andean colonists will move toward the same level of capitalization. In Upper Huallaga, the producers of legal crops face a series of difficulties. Market conditions and the pricing structure often impede the transformation toward an economy of commercial agriculture. While these poor peasants are not strictly guided by the principles of use values, neither do they follow a simple economic rationale of self-sufficiency. Rather, their subsistence strategies are structured around very restricted patterns of maximization. In quantitative terms, the level of consumption within this sector is extremely limited. Levels of accumulation are similarly low. These agriculturalists, striving to reach a specific level of consumption each year, which varies according to both the regional and national economies, come more and more to resemble settlers in the initial stages of colonization. They also begin to resemble the poorer, noncoca-producing population who practice a migratory agriculture far from roads and other commercial urban centers, a group with high annual rates of deforestation. Social differentiation accelerates in the colonized zones of the upper rain forest because of the specific conditions of articulation with the market economy.

Briefly stated, the relatively longer duration of the colonists in the settled regions has allowed a greater degree of capitalization among some of the farmers; with others the outcome at the end of a certain number of years has been unfavorable market relations. In each case the management of natural resources is different, but it always closely relates to the dynamic of the external economy.

Indigenous Peoples and Deforestation

Different ethnic groups in the lowlands manage natural resources differently. A study conducted between 1984 and 1985 by the Center for the Investigation and Promotion of the Amazon (CIPA 1986b) provides baseline comparative data for examining this issue. The recently published reports of anthropologists and geographers who have been conducting research in various regions of the Amazonian lowlands offer more information.

The purpose of the CIPA project, which involved approximately a dozen investigators, was to examine the impact of the market economy on the indigenous communities of Satipo, Lower Urubamba, and Madre de Dios.[2] The study included 94 family units from Satipo, 65 from Urubamba, and 67 from Madre de Dios.[3] The size of the samples was calculated according to the size of the population in each region. The indigenous population of Satipo occupies the upper rain forest zone of central Peru, the communities of Lower Urubamba are in the south-central Peruvian rain forests, and those of Madre de Dios are in the Amazonian lowlands.

The data collected during the CIPA study allows for a comparative analysis of these three regions. Ways in which the productive strategies of the native population differ from those of the colonists in the Satipo and Upper Huallaga will also be addressed.

Availability of Natural Resources and Deforestation

The formulas used to determine rates of deforestation associated with the colonists are not applicable to the indigenous populations, whose subsistence strategies vary with the amount of natural resources available to them. We determined the rate for the indigenous populations of the three regions by determining the farm area cleared each year to cover household subsistence needs. In spite of differences in resource management among the three, this provides a general comparison.

The Amarakaeris Indians of the Madre de Dios region exercise use rights over vast tracts of land for hunting, fishing, and plant resources. They usually manage three fields in different stages of the productive cycle: one in fallow, one at the peak of its output, and one in its second or third year of production (Moore 1985). Each field is isolated from the others and is under production for an average of three years, the period during which the annual crop yield is more or less optimal. After this period, the field is no longer used for annual crops, though it is maintained for such other types of crops as fruit. Given this production schedule, the Amarakaeris Indians typically open one small field per year. According to the CIPA (1986b) study, these fields average 0.31 hectares in size.[4] This represents the annual rate of deforestation.

According to Sioli (1980: 259–260), this technological system of shifting agriculture does not affect the continuity of the virgin forest in space and time. The temporary small clearings may locally interrupt the circulation of nutrients, but they "do not cut a great vein or main artery; they are only like needlepricks in the consistent forest cover, which soon heal again and do not leave permanent wounds" (ibid.). Moreover, since small clearings are surrounded by high virgin forest and deprived of protection only for short periods of time, soil erosion is not strong. The traditional practice of slash-and-burn shifting agriculture is within the limits of the buffering capacity of the tropical forest ecosystem (ibid.).

The natural resources available in the region of Madre de Dios enable the Amarakaeris to continue traditional hunting, fishing, and gathering, but these Indians have also become heavily involved with gold mining. This diversification plus the conservation ethic of the Amarakaeris (Moore 1985) have resulted in a notably slower rate of deforestation.

The Machiguenga of the Urubamba zone, like the Amarakaeris, have available a wide variety of natural resources, though relatively somewhat less abundant. The CIPA study indicates that the Machiguenga typically have two or three fields at different stages in the production cycle under cultivation.[5] The average size of these fields is 0.85 hectares. However, thanks to the development of cacao, a permanent crop that does not require the continual clearing of new land, the natives of the Urubamba region open one field annually, averaging 0.67 hectares, in order to plant manioc, corn, bananas, beans, and other food crops, and to maintain the agricultural cycle and cover its subsistence requirements (CIPA 1986b; see also Johnson 1983: 53–55). If we take into consideration that the Machiguenga have an average of 0.18 hectares of cocoa, and we assume that these fields are renewed every twelve years (because of their extensive type of cultivation) then the overall rate of deforestation is 0.68 hectares per year. This latter figure includes the clearing for annual and permanent crops. The differences observed in terms of average field size and annual deforestation rate with respect to the Amarakaeri of Madre de Dios reflect the fact that the indigenous peoples of the Urubamba region are more dependent on agriculture and less engaged in hunting, fishing, gathering, and mining.

Like the other two groups discussed above, the Ashaninka of Satipo typically utilize an average of three fields per family unit. The average size of these fields is 1.26 hectares.[6] In contrast to the regions of Urubamba and Madre de Dios, the Satipo area, because of the massive colonization of the past forty years, is being cleared of second growth, not virgin forest. The Ashaninka maintain, nevertheless, a number of coffee plantations, which reduces the annual rate of deforestation. The average number of coffee hectares is 0.50 and the average of annual crops is 0.76 hectares. Again, if we assume that the clearing of forests for temporary crops occurs every year, then we have an annual deforestation rate of 0.76 hectares. On the other hand, if we assume that the clearing of forests with the intent of developing plantations for permanent crops occurs only every ten years, then the overall deforestation rate—which entails clearing for temporary and perennial crops—is 0.81 hectares per year. The higher rate of deforestation noted for the Ashaninka of Satipo with respect to the other two ethnic groups reflects their almost exclusive dependence on agriculture and coffee production, for the Satipo region has much less in the way of natural resources than the other two zones.

This figure represents approximately 45 percent less than the current average rate calculated for the slash-and-burn agriculture practiced by colonists in the Upper Huallaga, which is estimated at 1.47 hectares per year. It is 57 percent lower than the global rate of deforestation—1.89 hectares—which includes fallow area, pastures, and agricultural crops, attributed to these same colonists. Likewise, the global rate of deforestation calculated for the colonists in the Satipo region is 2.13 hectares per year, whereas the amount of deforestation associated with the Ashaninka is 62 percent less than this figure (CIPA 1986a, 1986b). In other words, while the Ashaninka have a higher rate of deforestation with respect to the other indigenous Amazonian groups in the CIPA study, it is still considerably less than the rate of destruction attributable to the colonists located in the Upper Huallaga and Satipo.

In summary, the greater the availability of natural resources, the slower the rates of deforestation for the three ethnic groups under consideration. This lower rate of destruction correlates with a diversification of productive activities. Conversely, a greater dependence on agriculture to the exclusion of other types of productive activities leads to higher rates of deforestation. The relationship between natural resources and deforestation rates noted among the indigenous groups is the inverse of that described for the colonists in these regions. Among the colonists, the more land available, the greater the rates of consumption and destruction of the forest. For the native population, the more land available, the greater the number and types of natural resources available, and the more diversified the approach to the exploitation of these zones.

Slash-and-burn agriculture among the indigenous groups is radically different from the migratory agriculture of the colonists. It is the first phase of a prolonged system of agroforestry, as several authors confirm (Irvine 1989: 225; Clay 1988; Denevan et al. 1984; Denevan and Padoch 1987). The native populations do not merely restrict the farm's resources to two or three years of annual crop harvests, but maintain the fallow land in fruit trees and other trees of high commercial timber value or of value for medicine or construction. Moreover, much of the land that they clear is not exactly a natural forest, but a managed forest. These management strategies conserve the structure and composition of the tropical forest. Among the Andean colonists, such a resource management strategy does not exist, particularly for the fallow land in the tropical zones. All three indigenous populations, while differing from one another, show a lower rate of deforestation than any of the colonists, but the comparison between the two groups must not be limited to the size of the areas deforested, it should also look at resource management in general.

A controversial issue is the conservationist ethic that, according to Moore (1985), manifests itself in a rational system of natural resource management among the Amarakaeri. This argument emerges from Moore's extensive field experience with the Amarakaeri of Madre de Dios. Nevertheless, Johnson (1989), in a recent article on the Machiguenga of Urubamba, maintains that this group's judicious nondepleting use of natural resources owes itself to low population density and not to a deliberate conservationist goal. In his fieldwork Johnson did not encounter a conservationist ethic among the Machiguenga. Certainly, among some ethnic groups, like the Amarakaeri for example, the conservationist ethic may constitute an important part of their technological culture, but in other cases reduced demographic pressure on natural resources may determine their management. Likewise, Johnson argues that the Machiguenga approach to their environment reveals an "economizing calculation of self interest" (Johnson 1989: 220) by independent small groups that practice a family-centered technology. However, Johnson did not sufficiently analyze the higher or lower level of articulation of the Machiguenga to the market economy as a regulating factor in the rational or destructive use of the natural resources. If we take into account the economic relationship of the indigenous groups to the market, Johnson's argument concerning the effect of demographic pressure on natural resources cannot be discounted, but assumes other variables. The population factor is located in a larger context and is not the single element explaining the appropriate or inappropriate use of natural resources. In the next section we will continue developing this theme.

Economic Orientation and Deforestation: Indigenes and Colonists

Different economic rationales underlie the productive activities of each of the three native groups. The economic orientation of agriculture in each area was calculated from the volume of harvests in kilograms. The calculations were obtained after a three-to-four-month field study under the direction of anthropologists or field technicians who had prolonged experience in the regions being studied, in some cases more than five years. The indigenous people of Madre de Dios actively participated in the investigation. In that Madre de Dios region, which has the slowest rates of deforestation, 72 percent of the family productive units are predominantly oriented toward self-sufficiency, while another 24 percent focus exclusively on this goal (CIPA 1986b: 45). This means that 96 percent of nuclear and extended families in the Madre de Dios region organize their agricultural activities around the concept of family self-sufficiency (CIPA 1986b: 46–47). Among the Amarakaeri, who are heavily involved in gold mining, the average number of hectares under cultivation per domestic unit is 1.04, the main cultigens being manioc, corn, and bananas (CIPA 1986b: 39). The income from gold has led to a reduction in the amount of time devoted to agriculture in comparison to previous generations (CIPA 1986b).

One of the most remarkable aspects of the Amarakaeri's participation in gold extraction is the high income it generates, yet the results of the CIPA investigation indicate that this money has not disrupted the Amarakaeri's traditional productive organization, characterized by a variety of economic activities. Although they dedicate less time to traditional tasks, they have not abandoned hunting, fishing, or foraging, or modified their nondestructive management of resources. The greater availability of land has allowed the Amarakaeri to maintain their slash-and-burn agriculture, oriented to the unit of production and consumption, and they have continued it utilizing their own techniques characteristic of indigenous Amazonian peoples.

As already indicated, in the Lower Urubamba—which is occupied by the Machiguenga and the Piros—the deforestation rates are higher than in the Madre de Dios region. According to the CIPA report, the relative degree of self-sufficiency in this region is also high. The agricultural activities of 72 percent of the domestic units are primarily oriented toward the goal of economic autonomy, while 9 percent cultivate exclusively for family consumption (CIPA 1986b: 47). In other words, 81 percent of the production and consumption units included in the CIPA study utilized their crops to fulfill the dietary requirements of the household. In contrast to the Madre de Dios region, 19 percent of the Machiguengas and Piros of Urubamba market the majority of their agricultural produce. Even those who are primarily involved in production for domestic consumption sell a larger percentage of their harvest (CIPA 1986b: 45).

Likewise, as we have previously indicated, the rates of deforestation among the Ashaninka of Satipo are the highest noted for the three ethnic groups included in the study. A number of economic and social processes have contributed to this situation. In contrast to the other two regions, monetary income and cash crops have become more important for the Ashaninka than subsistence farming, and traditional productive activities have declined in significance (CIPA 1986b: 69). In the Satipo region, commercial agriculture centers around cacao and coffee, produced for external markets. This emphasis on one or two crops, together with the influx of peasants from the highlands, has put a great strain on the natural resources of the land. This has in turn led to changes in the logic of economic self-sufficiency associated with the more traditional patterns of the agriculture.

In Satipo, agricultural production is specialized to varying degrees. The degree of specialization appears to correlate with who the intended recipients of the produce are. While coffee, bananas, cacao, and papaya are commercial crops produced for export, such other items as avocados, rice, pitucas, and manioc are primarily for domestic consumption. An estimated two-thirds of the Ashaninka's productive units in the re-

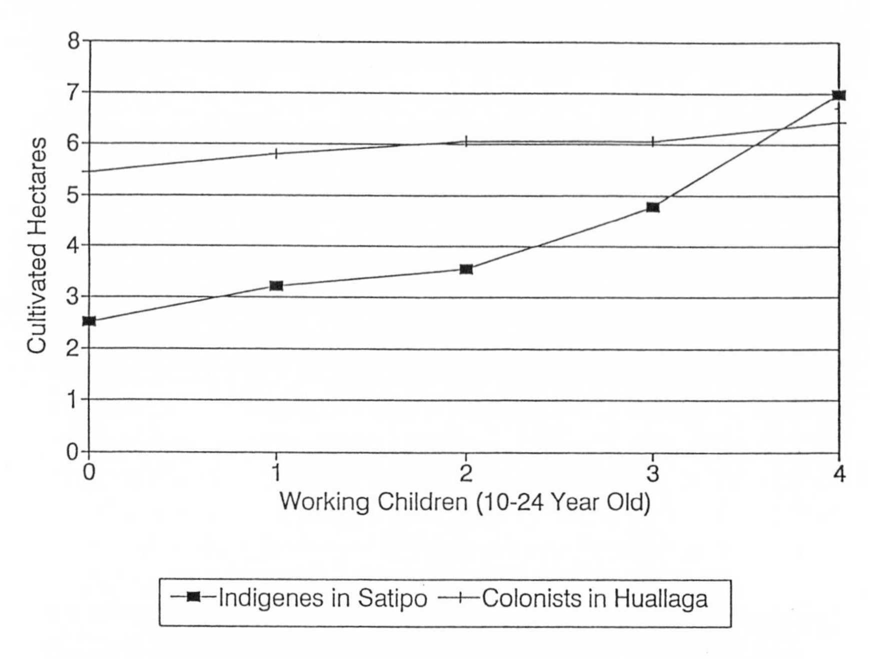

Fig. 2. Average cultivated hectares and family workers among colonists and indigenes. (Data from Survey FDN 1981; CIPA 1986b.)

gion produce for the market while the remainder produce primarily for domestic use.

The natives and the settlers also organize and utilize human resources differently, which implies a difference in the management of natural resources. The availability of family labor is more critical for the Ashaninka than for the colonists of Upper Huallaga. According to a study conducted by the Foundation for National Development (FDN 1981), the average increase in the amount of land under cultivation in the Upper Huallaga for each child of working age is 0.24 hectares. This is only 21 percent of the rate of increase in new agricultural land (1.11 hectares) that the Ashaninka experience with the entrance of a ten-year-old son into the productive life of the family. Even though the settlers work a larger number of hectares on average, the contribution of each son of working age is not as significant as in the native population. Figure 2 shows the importance of family labor among the Ashaninka in the strong correlation between the number of children of working age in a family and the number of hectares under cultivation.

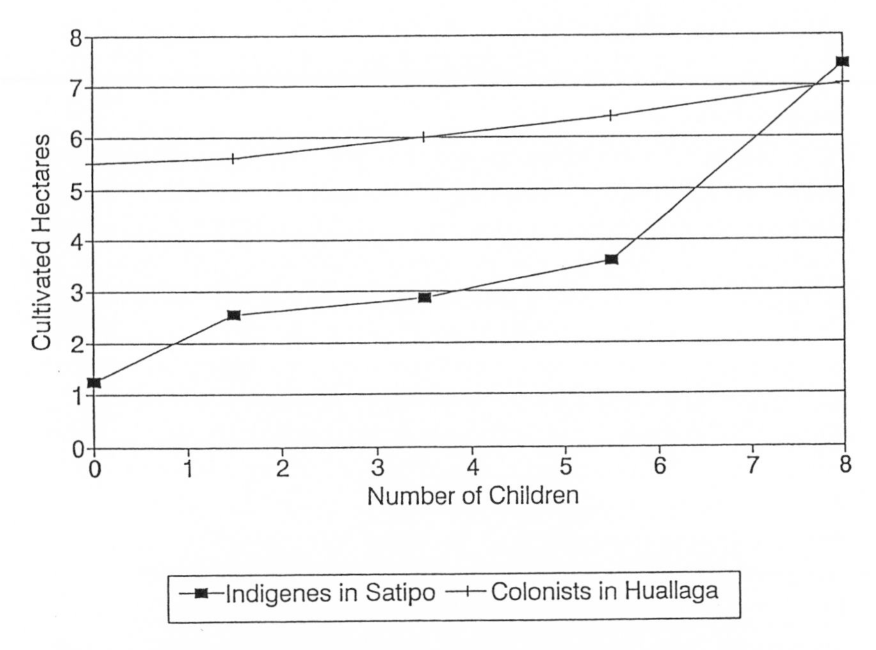

Fig. 3. Average cultivated hectares by number of children among colonists and indigenes. (Data from Survey FDN 1981; CIPA 1986b.)

Among the colonists in the Upper Huallaga, except in the initial stages of colonization, this correlation is not as significant. While the indigenous population depends heavily on the family labor force, the settlers combine family labor with wage labor to increase the amount of land in agricultural production (CIPA 1986b: 43–44). According to one study, 75 percent of the settlers in the Upper Huallaga region contracted seasonal wage labor (Aramburú 1984).

Natives also differ from settlers in the association between the total number of children and the amount of land under cultivation. In the Upper Huallaga, the correlation of these two variables is not as great as in the case for the Ashaninka Indians (fig. 3). The family size of colonists in the Upper Huallaga does influence the processes of agricultural intensification in zones of greater population density. In less heavily populated areas, it is the number of sons of working age that is most strongly correlated with the number of hectares under cultivation (Aramburú and Bedoya 1987). Family size never, however, has the same overall importance that it does for the indigenous communities.

The significant revenue that the Ashaninka Indians generate by cash crops, which they use for obtaining urban-produced foodstuffs, has reoriented their agricultural production to the market economy. Finding themselves involved in a regime of commercial production, they organize their production strategies accordingly, but, as the CIPA study concludes, their participation in the market economy is motivated by the need to satisfy the basic requirements of the domestic productive unit, not by the logic of accumulation. Once the biological and cultural requirements have been met, production declines (CIPA 1986b: 92).

In sum, the productive strategies of the indigenous populations that occupy the three regions discussed above operate within a logic of simple reproduction or of simple commodity production. By looking at the various factors involved more holistically, one learns why the indigenous peoples of the Peruvian Amazonia have strategies of resource management and rates of deforestation that differ significantly from those of recent settlers in the region.

Summary and Theoretical Implications

Agricultural systems, the specific technological methods by which a human society obtains edible crops, will always correspond to its forms of social and economic organizations. Specific economic rationales also result from a particular articulation of forms of production, such as urban capitalist economy with peasant economy. Likewise, agricultural systems also have a relation to certain levels of local and regional demographic density, and to the availability of natural resources (Turner and Brush 1987). Such a definition of agricultural systems leads us not to a study solely of social factors involved in a particular technology, but to an understanding that social and economic relations continuously affect technological decisions (Bettelheim 1972).

Among the colonists, the more land available, the greater the rates of destruction of the forest. In the case of the colonists in Upper Huallaga, however, we cannot limit the analysis of the intensification of agriculture and deforestation to a description of the characteristics of slash-and-burn agriculture and the availability of natural resources. The effects of the illegality of coca, its high production costs, and above all the high income it generates, provide eloquent proof that such structural factors as prices for agricultural products and their inputs are determinant variables and even more important than legal land deeding in the process of intensifying land use. The legal system may improve the conditions of production for farmers, but a change in land tenure laws alone is not sufficient to modify land use patterns radically.

Among the native groups under discussion is the Amarakaeri of the Madre de Dios region. The exploitation of gold mining on the part of the natives of this region does not signify an expropriation of ethnic territories, but rather has led to a partial modification of the productive strategies of the group. The search for gold has displaced such traditional productive labor as hunting, fishing, gathering, and subsistance agriculture, reducing the time dedicated to these pursuits in comparison to previous generations and redirecting human efforts towards the aforesaid economic activity. Nevertheless, because of the maintenance of great availability of resources within their territory, the native Amarakaeri reproduce the structural characteristics of native economy. Among such characteristics are the maintenance of a high economic diversification, and in the area of agriculture, a high number of crops oriented to family consumption. Clearly, this diversification is the cause of a low rate of deforestation. The penetration of capitalism has not produced a radical transformation of the conservationist management of natural resources, and thus, has not provoked an expropriation of the ethnic territory in this case.

At the other extreme, we have the Satipo region. In this zone, the immigration of the colonizing Andean peasants—especially since the fifties—has provoked a drastic reduction in the territories possessed by the native Ashaninka. So dramatic has been the reduction of natural resources for this ethnic group that altogether the colonists have on average more hectares in comparison to the natives. Consequently, the resources of hunting, fishing, and gathering have diminished to the point of almost disappearing. As a result, the native Ashaninka have placed more emphasis on agriculture; in particular, the spread of coffee plantations on their farms. It should be indicated, however, that in spite of the importance of commercial agriculture, the natives continue spending a large part of their time in subsistence agriculture. Nevertheless, in spite of being almost exclusively farmers, the Ashaninka continue to maintain a high number of crops per plot.

In contrast to the Amarakaeri, the natives of the Satipo region have had to face a different type of capitalist penetration. Their response has been to concentrate on agriculture; and for this reason, they have a rate of deforestation relatively higher than the Amarakaeri. The market economy has entailed a significant reduction of natural resources, which impedes economic diversification. In short, for the native population, the more land that is available to ethnic groups, the greater the number and types of natural resources available and the more diversified the approach may be to the exploitation of these regions. Also, the greater the availability of natural resources, the slower the rates of deforestation for the ethnic groups in consideration. In other words, the manner of capitalist penetration, and whether or not it produces an expropriation of natural resources, constitutes a very important element in our analysis. The regional population density or the availability of natural resources is a consequence of the type of expansion of the market economy.

NOTES

This chapter reports on work supported by the Institute for Development Anthropology and reports on work supported by Human Settlement and Natural Resources Systems Analysis (SARSA) Cooperative Agreement No. DAN 1135-A-00–4068–00, at Clark University and the Institute for Development Anthropology, funded by the U.S. Agency for International Development, Bureau for Science and Technology, Office of Rural and Institutional Development, Division of Rural and Regional Development. The views and interpretations herein are those of the author and should not be attributed to the Agency for International Development or to any individual acting on its behalf.

The author thanks Michael Painter, Sylvia Horowitz, and Jane Collins for their many helpful comments on different versions of this chapter, as well as Peter Little, Tammy Bray, Kim Munson, Lew Lama, and Monica Sella. Finally, I have to express my gratitude to Carlos Mora, CIPA Director, for his authorization to publish the results of the research on the market economy impact among natives in the Peruvian Amazon basin.

1. Soil use intensity is calculated by a formula that measures the relative weight of the fallow areas: Soil Use Intensity of a given plot = (number of hectares under permanent crops + number of hectares under annual crops) divided by (number of hectares under permanent crops + number of hectares under annual crops + number of hectares under fallow).

2. Several investigators participated in the CIPA study, among them Carlos Mora, Efraín Gonzales, Francisco Verdera, Carlos Aramburú, Thomas Moore, Luis Román, and Eduardo Fernandez. The composition of the final document was my responsibility.

3. The Satipo communities that were included in the investigation were: Yavirani, Santa Rosa de Panaquiari, Cushiviani, Río Berta, San Ramón de Pangoa, and Boca del Kitiari. The Urubamba communities included in the study

were: Bufeo Pozo, Pucani, Sensa, Puija, Carpintero, Miaría, Nueva Italia, Nueva Unión, Huao, Segaquiato, and Santa Clara. Finally, the ones from Madre de Dios: Villa Santiago, Santa Rosa, Boca del Inambari, Vuelta Grande, Barranco Chico, Puerto Luz, San Jose del Karene, and El Pilar.

4. In Madre de Dios, 92.82 percent of the households had between 0.1 and 0.5 hectare at the time of the study.

5. In Urubamba, 27.43 percent of the households had between 0.1 and 0.5 hectares, 39.43 percent between 0.51 and 1.0 hectares, and the remaining 33.14 percent more than 1.1 hectares.

6. In Satipo, 18.04 percent of the household units had between 0.1 and 0.5 hectares, 23.73 percent between 0.51 and 1.0 hectares, 33.23 percent between 1.1 and 1.5 hectares, and the remaining 25 percent more than 1.51 hectares. In other words, 58.23 percent had more than 1.1 hectares.

REFERENCES

Aramburú, C.
1984 Problemática Social en las Colonizaciones. *In* Población y Colonización en la Alta Amazonía Peruana. 65–79. Lima, Peru: Consejo Nacional de Población-Centro de Investigación y Promoción Amazónica (CNP-CIPA).

Aramburú, C., and E. Bedoya
1987 Poblamiento y Uso de los Recursos en la Amazonía Alta: El caso del Alto Huallaga. *In* Desarrollo Amazónico: Una Perspectiva Latinoamericana. 113–177. Lima, Peru: Centro de Investigación y Promoción Amazónica—Instituto Andino de Estudios en Población y Desarrollo (CIPA-INANDEP).

Bedoya, E.
1981 La Destrucción del Equilibrio Ecológico en las Cooperativas del Alto Huallaga. Serie Documentos No. 1. Lima, Peru: CIPA.
1987 Intensification and Degradation in the Agricultural Systems of the Peruvian Upper Jungle. *In* Lands at Risk in the Third World. P. D. Little and M. M Horowitz, eds. 290–315. Boulder, Colo.: Westview Press.
1990 Las Causas de la Deforestación en la Amazonía Peruana: un Problema Estructural. Working Paper No. 46. Binghamton, N.Y.: Institute for Development Anthropology.

Bedoya, E., and F. Verdera
1987 Estudio sobre Mano de Obra en el Alto Huallaga. Lima, Peru: Ronco Corporation.

Bettelheim, Ch.
1972 Cálculo Económico y Formas de Propiedad. Buenos Aires: Siglo XXI.

Boserup, E.
1984 Población y Cambio Tecnológico. Madrid, Spain: Gryalbo Editores.

CENCIRA
1973 Diagnóstico Socio-Económico de la Colonización Tingo María-Tocache-Campanilla. Lima, Peru: CENCIRA.
CIPA (Centro de Investigación y Promoción Amazónica)
1986a Informe de Estudio sobre la Leishmaniasis en la Selva Alta. Internal Document. Lima, Peru: World Health Organization.
1986b Impacto de la Economía de Mercado en las Comunidades Nativas de Satipo, Bajo Urubamba y Madre de Dios. Internal document. Lima, Peru: CIPA.
Clay, J. W.
1988 Indigenous Peoples and Tropical Forests: Models of Land Use and Management from Latin America. Cambridge, Mass.: Cultural Survival, Inc.
Collins, J.
1986 Smallholder Settlement of Tropical South America: The Social Causes of Ecological Destruction. Human Organization 45 (1):1–10.
Collins, J., and M. Painter
1986 Colonización y Deforestación en America Central: Discusión sobre Asuntos de Desarrollo. Working Paper No. 31. Binghamton, N.Y.: Institute for Development Anthropology and Clark University.
Denevan, W., and C. Padoch
1987 Swidden Agroforestry. New York: New York Botanical Garden.
Denevan, W. M., J. M. Treacy, J. B. Alcorn, C. Padoch, J. Denslow, and P. Flores
1984 Indigenous Agroforestry in the Peruvian Amazon: Bora Indian Management of Swidden Fallows. Interciencia 9 (6):346–357.
Dourojeanni, M.
1982 Posibilidades para un Desarrollo Más Integral en el Huallaga Central-Bajo Mayo. Boletín de Lima. No. 16–17. December. Lima, Peru.
1989 Impactos Ambientales de la Coca y la Producción de Cocaína en la Amazonía Peruana. *In* Pasta Básica de Cocaína. F. Leonard, and R. Castro de la Mata, eds. 281–299. Lima, Peru: Centro de Información y Educación del Abuso de Drogas (CEDRO).
Durham, F. Kathleen
1977 Expansion of Agricultural Settlement in the Peruvian Rainforest: The Role of the Market and the Role of the State. Paper presented to the Association of Latin American Studies and the Association of African Studies. 2–5 November. Houston, Tex.
ECONSULT
1986 Informe Final de la Evaluación del Proyecto. USAID no. 527-0244 Desarrollo Rural del Area del Alto Huallaga. Lima, Peru: ECONSULT.

FDN (Fundación para el Desarrollo Nacional)
1981 Plan de Ejecución del Proyecto de Desarrollo Rural Integral del Alto Huallaga. Social Report. Lima, Peru: FDN.
Foweraker, J.
1981 The Struggle for Land. Cambridge: Cambridge University Press.
Friedmann, H.
1980 Household Production and the National Economy: Concepts for the Analysis of Agrarian Formations. Journal of Peasant Studies 7 (2):158–184.
Hecht, S. B.
1984 Cattle Ranching in Amazonia: Political and Ecological Considerations. *In* Frontier Expansion in Amazonia. M. Schmink and Ch. Wood, eds. 366–398. Gainesville: University of Florida Press.
INE (Instituto Nacional de Estadística)
1987 Encuesta Nacional de Hogares Rurales. Lima, Peru: INE.
Irvine, D.
1989 Succession Management and Resource Distribution in an Amazonian Rainforest. Advances in Economic Botany 7:223–237.
Johnson, Allen
1983 Machiguenga Gardens. *In* Adaptive Responses of Native Amazonians. Raymond Hames and W. Vickers, eds. 29–63. Studies in Anthropology. New York: Academic Press.
1989 How the Machiguenga Manage Resources: Conservation on Exploitation or Nature? Advances in Economic Botany 7:213–222.
Martine, G.
1980 Recent Colonization Experiences in Brazil: Expectations versus Reality. *In* Land, People and Planning in Contemporary Amazonia. F. Barbira-Scazzocchio, ed. 80–94. Cambridge: Cambridge University Press.
Moore, Thomas
1985 Madre de Dios: Reseña Histórica. Internal Document. Lima, Peru: Centro de Investigación y Promoción Amazónica (CIPA).
Moran, E.
1984 Colonization in the Transamazon and Rondonia. *In* Frontier Expansion in Amazonia. M. Schmink and Ch. Wood, eds. 285–303. Gainesville: University of Florida Press.
Myers, N.
1980 The Primary Source: Tropical Forests and Our Future. New York: W. W. Norton and Co.
Painter, M.
1987 Intercambio Desigual: La Dinámica de Empobrecimiento de colonos y la Destruccion Ambiental en las Tierras Bajas de Bolivia. *In* Estrategias productivas y recursos naturales en la Amazonía, by Eduardo Bedoya, Jane Collins, and Michael Painter. 99–137. Lima, Peru: CIPA (Centro de Investigación y Promoción Amazonica).

Rasnake, R., and M. Painter
1989 Rural Development and Crop Substitution in Bolivia: USAID and the Chapare Regional Development Project. Working Paper no. 45. Binghamton, N.Y.: Institute for Development Anthropology.

Schmink, M., and C. H. Wood
1987 The "Political Ecology" of Amazonia. *In* Lands at Risk in the Third World. P. D. Little and M. M Horowitz, eds. 38–57. Boulder, Colo.: Westview Press.

Schuurman, F.
1980 Colonization Policy and Peasant Economy in the Amazon Basin. Bulletin of Latin American Studies and the Caribbean (Boletín de Estudios Latinoamericanos y del Caribe). December: 106–113.

Shoemaker, R.
1981 Colonization and Urbanization in Peru: Empirical and Theoretical Perspective. *In* New Approaches to the Study of Migration. D. Uzzel and D. Guillet, eds. Rice University Studies, Houston, Tex. 62 (2):163–175.

Sioli, H.
1980 Foreseeable Consequences of Development Schemes and Alternative Ideas. *In* Land, People and Planning in Contemporary Amazonia. F. Barbira-Scazzocchio, ed. Occasional Publication No. 3. Cambridge: Centre for Latin American Studies, University of Cambridge.

SIPA (Servicio de Investigación y Promoción Agraria)
1962 La Actividad Cafetalera en Tingo María. Servicio de Investigación y Promoción Agraria. Ministerio de Agricultura. Lima, Peru: SIPA.

Todaro, M.
1977 Economic Development in the Third World. London: Longman.

Turner, B., and S. Brush
1987 Comparative Farming Systems. New York: The Guilford Press.

Verdera, F.
1984 Estructura Productiva y Ocupacional en la Selva Alta. *In* Población y Colonización en la Alta Amazonía Peruana. 169–186. Lima, Peru: CNP-CIPA.

Conclusion

Chapter 7

Political Ecology and Environmental Destruction in Latin America

William H. Durham

Introduction: The Rise of Political Ecology

Among anthropologists interested in environmental issues, one finds today increasing dissatisfaction with the paradigms and principles of ecological anthropology as it developed before about 1980. What is now called the "old ecology" is criticized for neglecting the political dimensions of human/environment interactions, and thus for treating human communities as if they were fairly homogeneous, autonomous isolates, adapting—or sometimes failing to adapt—to a given exogenous environment. From the cultural ecology of Julian Steward (1955) through the systems ecology of Roy Rappaport (1967), and on to the cultural materialism of Marvin Harris (1979), the primary focus was upon mechanisms of population adjustment to the natural environment. The importance of political dynamics—both those internal to populations, as may be fundamental to differential access to resources within the aggregate, and those between local populations and the wider world—did emerge as a theme of a few specific works, such as Clifford Geertz's *Agricultural Involution* (1963) and John Cole and Eric Wolf's *Hidden Frontier* (1974), but these few case studies fell short of providing analytic tools worthy of a general new approach. A truly "political ecology" has only begun to emerge in the 1980s.

This small and seemingly innocuous piece of anthropological history has two consequences of particular significance today. The first concerns the nature of ecological inquiry within anthropology. As Michael Painter points out in the introduction to this volume, contemporary "environmental destruction and what to do about it" has not been taken up as a major research issue in the discipline. There are, of course, notable exceptions, including works by various individual authors (for example, Bodley 1988; 1990; Durham 1979; Moran 1981) and such groups as Cultural Survival, Inc. and the International Working Group on Indigenous Affairs (IWGIA). Still, the importance of the topic is not reflected in the research priorities of the discipline. The second, related consequence is that anthropological concerns, despite their intrinsic relevance, are rarely voiced in the contemporary debate over environmental issues. Instead, the terms of debate today over the "critical challenges" of population, resources, and environment (UNFPA 1991) are generally those laid down by plant and animal ecologists and economists. The environmental consequences, for example, of ethnicity and the cultural evaluation of resources, of the internal social structure of human populations, or of the more global structure of international relations, to name a few, are often ignored or downplayed. The concerns of anthropology and related social sciences are missing, just when they are most needed. One of the main reasons for this, or so it seems, is the apolitical analytical tradition of ecological anthropology.

The Environmental Impact of Human Populations

A simple but revealing example will help to make the point. One of the focal topics of environmental discussion today is the subject of human impact on the environment. Guiding much of this discussion, and serving as a springboard for exploring policy options, is an uncomplicated three-term heuristic called the "IPAT" equation, which is featured prominently in a number of influential books about the future of the planet (e.g., Ehrlich and Ehrlich 1990: chaps. 3 and 6; 1991: chap. 1; Meadows, Meadows, and Randers 1992: 100–103; UNFPA 1991: 16–21). My colleagues Paul Ehrlich and Anne Ehrlich summarize IPAT as follows (1990: 58):

> The impact of any human group on the environment can be usefully viewed as the product of three different factors. The first is the number of people. The second is some measure of the average person's consumption of resources (which is also an index of affluence). Finally, the product of these two factors—the population and its per-capita consumption—is multiplied by an index of the environmental disruptiveness of the technologies that produce the goods consumed. The last factor can also be viewed as the environmental

> impact per quantity of consumption. In short, Impact = Population × Affluence × Technology, or I = PAT.

As a statistical summation, of course, there is no problem: as long as the averages are correctly defined and adequately measured, I must equal PAT. But as an attempt to "represent the processes involved," as it is often portrayed (e.g., UNFPA 1991: 17), I = PAT is worse than misleading. The message goes out that, to a first approximation, one need not bother with the internal structure of human populations (including ethnicity, gender, class, power relations, etc.), with internal cultural differences in resource use and technology, or with the surrounding world system of interpopulational relations. In effect, the message is that anthropological concerns—not to mention those of other social sciences—can be left out of the analysis. Not surprisingly, that is precisely what happens.

These shortcomings of IPAT become particularly severe, of course, at high levels of social aggregation, as when the equation is applied to some multi-ethnic region or to a whole nation-state. Ironically, however, it is at this level that IPAT is most often used. At first glance, such uses have a lot of appeal and may seem almost intuitive. For example, I find it easy to believe that "Because of this combination of a huge population, great affluence, and damaging technologies, the United States has the largest impact of any nation on Earth's fragile environments and limited resources" (Ehrlich and Ehrlich 1991:9). But can this really pass for an understanding of the processes behind our impact on environment?

The issues may be easier to see in Latin America. Consider Panama, for example, where human impact in the form of accelerating deforestation stands as a daunting challenge to contemporary conservation efforts. It is tempting to turn to IPAT and assume that the process of deforestation can be represented by the product of Panama's population (now over 2.5 million) times its affluence times its technology. But consider the problems raised by this assumption. First, most of the forest in Panama has been converted to pasture to feed an exponentially growing herd of cattle (see, e.g., Heckadon and McKay 1984:24; Joly 1989), a good portion of which, historically, has not been "consumed" in any meaningful sense within Panama, but exported to the world market. Moreover, there are whole populations within Panama—like the indigenous Kuna, for instance, with a population size of over 50,000—who not only produce no cattle but actively and effectively oppose cattle production (see, e.g., Breslin and Chapin 1984; Archibold 1992). Who then should be counted in Panama's P? Should the Kuna and other non-consumers be left out? But how can they be left out, when efforts by the Kuna and other indigenous groups to rid their homelands of outsiders and cattle have also played a key role in setting the A values of the rest of the country? The point is that all this is hidden, not revealed, by IPAT. To make matters worse, IPAT fails to convey the fact that the *implied* A for the country (that is, the total forest converted to cattle per Panamanian, for example) is not the *actual* A for Panamanians. Though it may well look like local Panamanian "affluence," much of the actual consumption of the cattle takes place far from the region. Because IPAT misses these important connections, surely it cannot be taken to "represent the process" of human impact upon the tropical forests of Panama. In this and so many other cases, I hinges upon existing social relations within and between populations. The problem is that social relations are simply not well represented by the product of P times A times T.

Social Causes of Environmental Destruction in Latin America

The authors of contributions to this volume would all seem to agree that IPAT does not reach to the root of the problem. The contexts of their studies are all different; their analytical foci and tools of analysis are fairly diverse; and the conclusions of the chapters are often narrowly focused and quite specific. Yet on that one point there is general agreement: the leading causes of environmental destruction in Latin America are not represented by IPAT. Too much is left out to salvage the equation; a different kind of approach is called for.

The alternative message that goes out from these contributions is that the impact of human populations upon environments is mediated by cultural and political economic forces that do not act as simple multipliers and multiplicands, and cannot even be approximated in that way. Among such forces, it is argued here, are the social relations within and between populations, whose institutionalized form in Latin America essentially guarantees inequitable access to resources. Environmental destruction follows, so it is argued, from this basic inequality along two separate pathways, one that can be called capital accumulation, following the lead of the papers by Stonich and Painter, and another that can be called simply impoverishment.

Figure 1 offers a schematic summary of these two interrelated path-

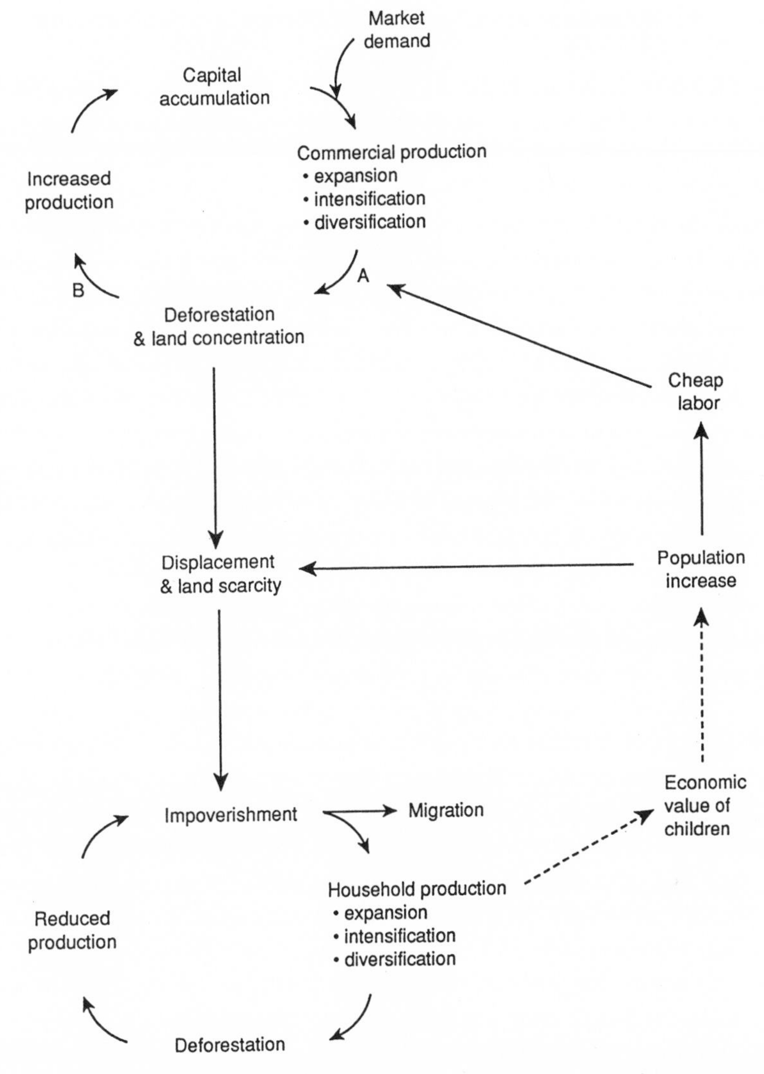

Fig. 1. The political ecology of deforestation in Latin America. This highly simplified sketch is an attempt to synthesize the main arguments of this volume concerning structural causes of environmental destruction in Latin America. The figure emphasizes two positive feedback loops—one corresponding to capital accumulation and the other to impoverishment—that promote deforestation, a form of environmental destruction discussed in all of the chapters. The loops are linked together by interdependent effects on population, resources, and environment. Dashed arrows indicate relationships suspected but not investigated in this volume. "A" refers to low input costs; "B" to production subsidies from nature and the state. The figure assumes market involvement and associated cultural values throughout the population(s) represented.

ways to environmental destruction in Latin America. Highly simplified to begin with, the schema focuses on the particular example of deforestation because it is a common concern among all of the chapters in this collection. (Note, however, that arguments in this volume imply that similar figures can be constructed for other forms of environmental destruction, or, indeed, for its less drastic form, degradation. For other attempts to illustrate such linkages see Collins 1986: 5–7; Leonard 1989: 7; Kates and Haarman 1992: 9.) The top feedback loop represents capital accumulation: fueled by domestic and/or foreign demand and aided by land laws, timber concessions, and the like, commercial production schemes expand into the remaining forest areas of a given region. With the advantage of what Edelman and others term "subsidies from nature and subsidies from the state," these production schemes (which may include ranching, logging, export crop production, or some combination) often generate lucrative short-term revenue in this way, a portion of which can then be reinvested in further expansion. When conditions are favorable, there is a positive feedback effect: successful deforestation produces funds that fuel its own acceleration. In addition, depending on land laws and tenure arrangements, the ensuing concentration of land and/or the displacement of previous occupants may also accelerate as a by-product of these dynamics.

In the Latin American context, the bottom feedback loop, impoverishment, is intrinsically linked to the top. The linkage occurs through the accelerating scarcity of land for household agricultural production and other uses, through the displacement of forest inhabitants from their homelands or, not infrequently, through both of these changes at once. It carries the inevitable consequence of lowered standard of living and/or reduced cash income for the affected peoples, which, in turn, may motivate various kinds of responses. For example, domestic units in the affected population may attempt compensatory increases in household production, through (a) expansion into new and presumably ever-more-marginal lands, (b) intensification of existing production via additional labor or other inputs, and/or (c) diversification of production so as to include more cash crops for income. But sooner or later, each of these efforts is likely to entail some form of environmental degradation, such as loss of soil and soil fertility, buildup of pesticide residues, and further deforestation. Frequently, all of this takes place in regions well removed from, and seemingly unconnected to (but only seemingly), the commercially driven deforestation of the top loop. In these circumstances, the subsidies from nature and the state are inevitably smaller, and the result is often ever-declining yields, further impoverishment of domestic units, and additional efforts at compensatory production, leading to further deforestation and degradation, and so on—what Painter calls, in chapter 4, the "mutually reinforcing cycle" of poverty and environmental destruction.

Given that there are two loops in the basic schema, one can always debate which is the more responsible for environmental destruction in any particular context, or indeed in a general sense. But an important point of the chapters in this volume is that such debate is a rather academic exercise: the two are intrinsically linked. They are two sides of the same coin, as it were; paired consequences of the same structure. In the figure, this linkage is represented by arrows between the loops: the displacement and land scarcity created by the top loop fuels impoverishment in the bottom, while at the same time producing a pool of inexpensive labor—both directly via displacement and indirectly via population growth and the economic value of children (a link not fully explored in this volume)—that is, itself, crucial to positive feedback effects at the top. Population increase is accorded an integral role in the scheme, and technology and affluence/consumption are implied within each of the two loops. In fact, differences in technology and consumption are among the reasons that the loops are distinguished and separated from one another in the first place. But these variables are imbedded in a full system of linkages and feedbacks that simply are not represented or even reflected in the I = PAT equation.

One final point of the arguments summarized by Figure 1 warrants mention. Among the possible responses to impoverishment in the bottom loop is migration—migration either to places characterized by newer and more promising loops of the kind shown here, or, indeed, to some "frontier area" not already fully integrated into such a structure of relations. The latter is of particular interest: although it is conceivable that migration could take individuals and groups to some place outside the reach of the loops and links discussed here, that is by no means a guaranteed result. On the assumption that frontier areas are generally still more marginal for the pertinent production systems (or they would already have been incorporated), and that migrants will still need goods, services, or income from the sending region, a more likely result is an eventual reproduction of the pattern in the frontier area, particularly the cycle of impoverishment and degradation. Indeed, it is not unrealistic to imagine an entire chain of lower loops, in ever-more-marginal habitats, spinning off via migration from the bottom loop of figure 1.

The Case Studies of This Volume

Although the contributions to this volume do seem to share key components of this alternative model for human impact, they also emphasize a number of different properties and features. Consider, first, Edelman's chapter on the "hamburger thesis." The chapter warns against the simplistic "dependency" version of these relations, wherein the trigger that sets the top loop in motion, and from there the full system, is simply foreign demand for a given export commodity. Instead, suggests Edelman, the top loop operates in a broader, more encompassing way, with foreign demand for a given export (or range of exports) being but one of the originating and sustaining conditions. Consequently, there is a good measure of resilience within the top loop, particularly when the subsidies from nature are large: parameters can change and substitutions can take place—for example, domestic demand can effectively take over for foreign demand—and still the cycle spins forward intact. In fact, the Costa Rican example indicates that commercial production of a key commodity in the loop, such as beef, can even be assaulted by a crisis of domestic and international origins, and still the loop can operate. After all, Edelman notes, the "natural capital" of the forest can subsidize any number of activities, not just cattle ranching, and it would be folly to assume that grass, beef, or any other particular commodity, "rather than money," drives the system forward. Adding to the persuasive force of Edelman's argument is an appropriate and skillful use of statistics from the FAO, the USDA, the World Bank, and Costa Rican agencies. One is left with little doubt that the loop at the top of Figure 1 is far larger than "the hamburger connection."

In further support for Edelman's point, the chapter here by Susan Stonich shows just how little it seems to matter what commodity or commodities are featured in the top loop. In the case of Honduras, Stonich argues that the outcome is always similar—according to measures of land scarcity and price, legal and illegal displacement of rural inhabitants, and indices of poverty, environmental degradation, and the like—whether the top-loop commodities are one or more traditional exports (bananas and coffee in the Honduran context), the nontraditional exports of one generation (e.g., cotton and cattle), or the nontraditionals of the next generation (e.g., melons and shrimp). Commodities come and go, Stonich notes, but the structure of the underlying relations and processes remains the same. Furthermore, the pattern holds even for commodities expressly introduced through aid programs to benefit the small, domestic producers. Because the two-loop system is already in place, what begins as a novel production supplement for smallholders quickly becomes another good investment opportunity for large-scale producers. With no exceptions to-date, she notes, the consequences "are an escalation in the rate of land concentration" and the "concomitant elimination of independent small and medium producers." The saga of southern Honduras, amply documented here and in other studies by Stonich, Billie DeWalt, Douglas Murray, and others, is among the more convincing on record that these dynamics are neither weak, intermittent, nor ephemeral, but built right in to the structure of the society. Perhaps hope will come from new players in the game: Stonich points out that the recent expansion of shrimp farms into Honduran mangrove forests occurred over the protests of various environmental groups. But it remains to be seen whether even the organized pressure of environmental concern can effectively override the ever-potent feedback effects summarized in figure 1.

The chapter by Norman Schwartz on deforestation in the Petén region of Guatemala effectively subjects the political ecology model to a different, and demanding, kind of test. As Schwartz notes, the Petén was officially opened to colonization and development efforts only after the creation of the oversight agency FYDEP in 1958; the question is thus, How well does the model apply to an area where the loops have really had only a few decades to develop? Do the same forces operate and do they produce the anticipated results? The answer, says Schwartz, is an unqualified "yes." As he aptly puts it, logging, ranching, road construction, and smallholder farming all "feed on each other. . . . Thousands of *milperos,* ranchers, loggers, and speculators not only compete for relatively easily exhausted land, but each also paves the way for or forces the other to clear more and more forest." The linkages are particularly clear in the case of late-arriving *milperos*: after clearing remote, small plots with poor soils, they have no choice but to join the seasonal pool of cheap labor, which in turn "helps make logging (and other enterprises) profitable and encourages its expansion," thus opening new roads and ever-more-remote areas to milpero colonization. Among the strengths of the chapter are its encompassing treatment of the multiple sectors of the Petén economy, and its exposure of the "overdeterminacy" in the deforestation loops. On the other hand, the presentation is largely descriptive, leaving one eager to see full analytic support of its claims. Still, the message comes though clearly: "deforestation serves the immediate interests of so many people and institutions that it will prove difficult to halt, short of turning the entire country around," and all of the loops with it.

In many ways, the chapter on "Upland-Lowland Production Linkages" by Michael Painter exemplifies the political ecology approach, and, indeed, it provided critical inspiration and guidance in my attempt to synthesize the big picture in figure 1. In addition to the paper's commendable logic and exposition, there are two further features worthy of special mention. The first concerns the careful elucidation of pathways and interconnections, particularly those governing household production in the lower loop. Painter takes considerable pains to point out just how impoverishment and deforestation are interrelated. In the case of livestock management, for example, he describes multiple pathways: one runs from decline of rural incomes to high rates of male outmigration to household labor scarcity to deterioration of livestock management to the degradation of grazing lands and on to the wastage of useful manure. Another runs a parallel course from (a) increasing economic hardship, to both (b) encroachment of agriculture on grazing lands and (c) a steady increase in herd sizes, resulting in (d) higher intensity of land use for grazing and crop production, in turn producing (e) degradation in the form of erosion and decline in soil fertility, thus feeding back into (a). Although multiple reinforcing pathways greatly complicate the problems of data collection and analysis, they also make the linkages stronger and more convincing. In the Bolivian context, they underscore Painter's point that, since most migrants are men, "the principal agents and victims of land degradation are women and small children." In a society with heavy gender-based discrimination, this adds further complexity to the task of breaking the cycle of poverty and degradation.

Speaking of data analysis, a second laudable feature of Painter's chapter is his statistical documentation and hypothesis testing. Thanks to data sets assembled in collaboration with Bolivian agencies, including one sample with a whopping *N* of 10,703, Painter is able to pose and test answers to a number of questions underlying his analysis. For example, in answer to the question of why Chapare farmers grow only limited amounts of coca, Painter uses these data to show that coca plantations are limited neither by access to land nor by availability of family labor. His alternative suggestion—that the price of coca is "volatile and often low," and thus "not sufficiently profitable to provide incentives to hire labor"—is certainly strengthened, if not fully substantiated, by the findings to date. Nevertheless, a crucial point has been made: if political ecology is to gain the credibility and influence it deserves, rigorous tests of key hypotheses are an important way to go about it.

The chapter by James Jones on the Beni Department of Bolivia, just north of the Chapare, complements Painter's contribution in both its historical and interethnic dimensions. With respect to history, Jones documents the actors and events responsible for bringing the two-loop system into the Beni. He shows that crucial components were fully established in the region only quite recently, in the postwar "dynamic period." And even then, changes came slowly at first, largely via the commercialization of cattle and the titling of lands after the Agrarian Reform of 1953. By 1965, however, with soaring beef prices and the promise of new, government-built roads, the region witnessed a veritable "melee of land rush and land reform," complete with redefined laws of land tenure, from which the Beni's indigenous population emerged "landless and destitute" by 1979. By this point, even the peltry trade took on the dynamics of the lower loop: "As Indians grew poorer, they hunted more, struggling always to reduce mounting debt with peltry," and moved ever deeper into degrading forests, "where they depended on traders who demanded payment in peltry."

But if the system came to the Beni only recently, it has certainly been no less effective. As Jones's paper emphasizes, the consequences of poverty and environmental destruction were especially acute among various Indian populations in the region, including Chimán, Sirionó, Yuracaré, and Trinitario. As if to prove his point, at the height of Bolivia's "worst economic crisis of the century," the government went so far as to grant seven commercial timber concessions in areas that were, by this point, the Indians' "last refuge." The widely publicized "long march" of Beni Indians in 1990, the concessionary (if short-lived) executive orders of the president, subsequent indigenous rights legislation, and the recent declaration of a five-year "Ecological Pause" all add up to a kind of national admission of the devastation wrought by the two-loop system in the context of the Beni. Although I find myself wishing that Jones had imparted a stronger theoretical framework to his chapter, including lucid exposition of its testable hypotheses, on the general point he is surely right: the Oriente is a kind of "window" on Bolivia's past. Through it, one can see clearly the feedback loops and linkages that have driven degradation, poverty, and the marginalization of indigenous peoples.

The final chapter of the section, Eduardo Bedoya's analysis of the social and economic causes of deforestation in the Peruvian Amazon, reads, in many respects, like a methodological and thematic integration of the previous two. Like Painter, Bedoya's interests include colonists who have emigrated from the Andean highlands in the hopes of gaining improved access to land. Moreover, conditions in the Upper Huallaga region, where many of the colonists have settled, are not unlike conditions in the Chapare; for example, one finds similar conventional crops (like maize, rice, bananas, coffee, and cacao), not to mention coca, whose explosive growth in this region of Peru since the mid-1970s is almost legendary. Furthermore, Bedoya is also able to put to good use some fairly recent survey data collected in collaboration with Peruvian agencies (in this case, the Foundation for National Development and the Center for the Investigation and Promotion of the Amazon). Among other findings, he confirms that colonist agriculture is generally quite extensive in the Upper Huallaga—meaning that deforestation rates are relatively high—because of a pervasive labor shortage that prevents intensification. The trend is particularly pronounced in the case of squatters, whose dependence upon annual crops fuels an average annual deforestation rate of 2.72 hectares, more than double that of legal tenants (1.28 hectares; unfortunately, Bedoya's tables do not include averages across all categories). But the reason why labor is scarce in the zone, Bedoya affirms, is that coca producers offer wages at least 125 percent higher than non-coca producers. What he terms "legitimate farmers" are thus locked into a cycle of relative poverty and high rates of annual clearing.

But Bedoya's chapter also includes themes in common with Jones's analysis, particularly with respect to the impact of the encroaching market economy on the indigenous peoples of the Amazon. In this respect, Bedoya is able to provide a comparative, quantitative test of a key hypothesis. If the political ecology model is correct, the indigenous peoples who are more integrated into the market economy, and thus the loops of figure 1, should—other things being equal—clear more forest area per household per year than those who are less market oriented.

Drawing on data from the CIPA collaboration, Bedoya thus compares household clearing rates for a sample from an Amarakaeri community in the lowlands near Madre de Dios with those from a Machiguenga community on the Lower Urubamba River and those from a Ashaninka community near Satipo in the upper rain forest zone. If one then assumes roughly comparable household sizes and crop yields at the three sites (an assumption not adequately discussed in the chapter), the lowest rates of annual clearing are expected for the Amarakaeri, among whom 96 percent of the households "organize their agricultural activities around the concept of family self-sufficiency" (supplemented with income, it must be emphasized, from active gold mining); intermediate rates are expected among the Machiguenga, with a self-sufficiency figure of 81 percent; and the highest rates are expected among the Ashaninka, with a self-sufficiency estimate of 33 percent. According to Bedoya, this is precisely what one finds in their data: the average figures are 0.31 hectares cleared per household per year by the Amarakaeri, 0.68 by the Machiguenga, and 0.81 for the Ashaninka. Certainly there is more work to be done, both to corroborate Bedoya's provocative findings and to test their validity in other contexts. Still, there is preliminary support for the model in these findings.

Finally, Bedoya's discussion also builds upon an important theme that Jones calls "culture's enduring influence." Drawing again on the CIPA data, Bedoya demonstrates what anthropologists have long argued—that cultural values play a profound role in shaping patterns of natural resource usage. The Ashaninka, to take the author's case-in-point, are clearly involved in the market economy through their production and sale of cacao and coffee. But unlike colonists from the outside, their participation is also "motivated by the need to satisfy the basic requirements of the domestic productive unit, not by the logic of accumulation . . . Once [their] biological and cultural requirements have been met, production declines." In other words, the cultural values of the Ashaninka cut short the positive feedback of capital accumulation, thus keeping their economic system at least partially disengaged from the loops of figure 1. Moreover, the same appears to be true of the Machiguenga and Amarakaeri, but not of the colonists, even those in the most remote and isolated regions. The importance of the point is clear: Bedoya implies that cultural differences of just this kind explain why deforestation rates decline with aggregate land availability among the indigenous populations, while showing precisely the inverse trend in zones inhabited by colonists. If this analysis is any indication, political ecology would do well to pay increased attention to the role of culture in the causes of, and responses to, environmental degradation (on this point, see also Blaikie and Brookfield 1987; Charnley 1991).

Conclusion

The chapters in this collection offer what might be called a kaleidoscopic image of the social causes of environmental destruction in Latin America. On the one hand, the settings, the actors, the histories, the institutions, and the environmental problems, all inevitably specific to particular contexts, give rise to an almost bewildering collage when one looks upon the whole at once. On the other hand, careful scrutiny reveals an underlying symmetry among the chapters—a symmetry of structural origin—that shows itself as reflections or regularities in the arrangement of specific components. As argued here, such symmetry is the product of a system of relations that sets up two positive feedback loops, one characterized by capital accumulation and environmental degradation and the other by impoverishment and degradation. The loops are linked together by their interdependent effects upon such variables as resource scarcity, demographic increase, and population distribution. The chapters of this volume argue that this overall system of relations produces a set of powerful structural forces with nonlinear feedback effects, whose overall action shapes certain profound similarities despite very different contexts. The image is complex and multifaceted, yet also regular and comprehensible. Social relations are an important cause of environmental destruction in Latin America.

But there is still much to be done. A key implication of these papers is that further research is warranted both to expand the theoretical framework of political ecology and to test its applicability in particular contexts. For example, the model represented in figure 1 is clearly oversimplified. It dichotomizes a much more complicated social structure; it assumes Western market values in all sectors of society; and it leaves out perhaps even more than it includes. Among other things, we continue to need a far better understanding of feedback loops and their linkages, and more detailed assessments of their historical origins. We need far more attention to cultural variables in the causes of and responses to degradation. And we need far better ways to bring inequality in all its guises—race, class, gender, and ethnicity—into the picture. Still, some progress has here been made. In the kaleidoscopic studies of this volume, we have a fitting illustration of a new kind of inquiry within ecological anthropology and a glimpse at the promise of a genuine political ecology.

NOTE

For discussion, suggestions, and advice on the subject of this chapter, I thank Dominique Irvine, Michael Painter, Suzana Sawyer, Susan Charnley, Kathleen Foote Durham, and Stanford students in Anthropology 164/Human Biology 134, "The Human Ecology of Amazonia" (Spring 1991), and Anthropology 269, "Political Ecology" (Autumn 1992).

REFERENCES

Archibold, G.
1992 PEMASKY in Kuna Yala. *In* Toward a Green Central America: Integrating Conservation and Development. V. Barzetti and Y. Rovinski, eds. 21–33. West Hartford, Conn.: Kumarian Press.

Blaikie, P., and H. Brookfield, eds.
1987 Land Degradation and Society. New York: Methuen and Co.

Bodley, J. H., ed.
1988 Tribal Peoples and Development Issues: A Global Overview. Mountain View, Calif.: Mayfield Publishing Co.

Bodley, J. H.
1990 Victims of Progress. 3d ed. Mountain View, Calif.: Mayfield Publishing Co.

Breslin, P., and M. Chapin
1984 Conservation Kuna-style. Grassroots Development 8 (2):26–35.

Charnley, S.
1991 Cultural Perceptions of Ecological Change: The Case of "Degradation" in Usangu, Tanzania. Paper presented at the Third Stanford Centennial Symposium, "Ethnicity, Equity, and Environment," April 11–14, 1991. Stanford, Calif.: Stanford Alumni Association. Video.

Cole, J. W., and E. R. Wolf
1974 The Hidden Frontier: Ecology and Ethnicity in an Alpine Valley. New York: Academic Press.

Collins, J. L.
1986 Smallholder Settlement of Tropical South America: The Social Causes of Ecological Destruction. Human Organization 45 (1): 1–10.

Durham, W. H.
1979 Scarcity and Survival in Central America: Ecological Origins of the Soccer War. Stanford, Calif.: Stanford University Press.

Ehrlich, P. R., and A. H. Ehrlich
1990 The Population Explosion. New York: Simon and Schuster.
1991 Healing the Planet: Strategies for Resolving the Environmental Crisis. Reading, Mass.: Addison-Wesley.

Geertz, C.
1963 Agricultural Involution: The Process of Ecological Change in Indonesia. Berkeley: University of California Press.

Harris, M.
1979 Cultural Materialism: The Struggle for a Science of Culture. New York: Random House.

Heckadon Moreno, S. and A. McKay, eds.
1984 Colonización y Destrucción de Bosques en Panamá: Ensayos Sobre un Grave Problema Ecológico. Panamá: Asociación Panameña de Antropología.

Joly, L. G.
1989 The Conversion of Rain Forests to Pastures in Panama. *In* The Human Ecology of Tropical Land Settlement in Latin America. D. A. Schumann and W. L. Partridge, eds. 86–130. Boulder, Colo.: Westview Press.

Kates, R. W. and V. Haarmann
1992 Where the Poor Live: Are the Assumptions Correct? Environment 34 (4):4–11, 25–28.

Leonard, H. J. ed.
1989 Environment and the Poor: Development Strategies for a Common Agenda. New Brunswick, N.J.: Transaction Books.

Meadows, D. H., D. L. Meadows, and J. Randers
1992 Beyond the Limits: Global Collapse or a Sustainable Future. London: Earthscan.

Moran, E.
1981 Developing the Amazon. Bloomington: Indiana University Press.

Rappaport, R. A.
1967 Pigs for the Ancestors: Ritual in the Ecology of a New Guinea People. New Haven, Conn.: Yale University Press.

Steward, J.
1955 Theory of Culture Change: The Methodology of Multilinear Evolution. Urbana: University of Illinois Press.

UNFPA (United Nations Population Fund)
1991 Population, Resources, and the Environment: The Critical Challenges. New York: United Nations.

Index